REGENERATION OF THE COALFIELD AREAS

REGENERATION OF THE COALFIELD AREAS

ANGLO-GERMAN PERSPECTIVES

Edited by

Chas Critcher, Klaus Schubert and David Waddington

PINTER

in association with the
Anglo-German Foundation for the Study of Industrial Society
London and New York

PINTER
A Cassell imprint
Wellington House, 125 Strand, London WC2R 0BB, England

First published in 1995

British Library Cataloguing in Publication Data

A CIP catalogue record for this book is available from the British Library

ISBN 1 85567 205 7 (hb)

Library of Congress Cataloging-in-Publication Data

Regeneration of the coalfield areas: Anglo-German perspectives /
   edited by Chas Critcher, Klaus Schubert, and David Waddington.
      p.   cm.
   Includes bibliographical references and index.
   ISBN 1-85567-205-7
    1. Coal trade—Government policy—Great Britain—Congresses. 2. Coal trade—
Government policy—Germany—Congresses. 3. Mine closures—Great
Britain—Congresses. 4. Mine closures—Germany—Congresses. I. Critcher, C. II.
Schubert, Klaus, 1951– . III. Waddington, David P. IV. Anglo-German Foundation for
the Study of Industrial Society.
HD9551.6.R43 1995
338.2'724'0941—dc20   95-3375
                                                        CIP

Typeset by Saxon Graphics Ltd, Derby
Printed and bound in Great Britain by Biddles Ltd, Guildford and King's Lynn

CONTENTS

NOTES ON CONTRIBUTORS

Dr Heiko Beenken is Director of the Regional Structural Change Department at the Ministry for Employment, Health and Welfare of North-Rhine Westphalia.

Martin Bräutigam is a Research Assistant, Ruhr University, Bochum.

Dr Rainer Buhr is Director of Economic Development Agency. (Gesellschaft fur Wirtschaftsforderung mbH).

Chas Critcher is a Reader in Communication, Head of the Communication Media and Communities Research Centre, Sheffield Hallam University.

Dr Hans Estermann is at Developing Agency East Ruhr, Bergkamen.

Professor Hywel Francis is Professor of Education, University College, Swansea.

Franzis Roxlau-Hennemann is at Developing Agency East Ruhr, Bergkamen.

Dr Matthias Hessling is Head of Department, World Energy Markets/European Union, Ruhrkohle AG, Essen.

Professor Ray Hudson is Professor of Geography, Geography Department, Durham University.

Wulf Noll is Head of Section, Basic Questions of Structural Development, Ministry of Economics and Technology of North-Rhine Westphalia.

Ewart Parkinson is Planning Consultant, Associate Director, WS Atkins Wales.

David Parry is a Research Assistant, Communication, Media and Communities Research Centre, Sheffield Hallam University.

David Pickering is Commercial Director, British Coal Enterprise.

Mrs Pat Richards is Chief Executive, North Nottinghamshire Training and Enterprise Council.

David Sadler is a Lecturer in Geography, Geography Department, Durham University.

Hedley Salt is Leader of Barnsley Metropolitan Borough Council, President of The Dearne Valley Partnership, Chair of the Coalfield Communities Campaign.

Roland Schäfer is Head of City Administration, Bergkamen.

Dr Klaus Schubert is Assistant Professor, Gerhard-Mercator University, Duisberg.

Wolfgang Steingräber is Director of the Campaign for the Future of Coal Mining Areas, Hamm. (Zukunftsaktion Kohlegebiete (ZAK) e.v.).

Dr David Waddington is Senior Lecturer, Communication Studies, Sheffield Hallam University.

Friedrich Wilhelm Wagner is Mining Director, Section of Mining Planning, Deposit Securing and Mine Survey, Ministry of Economics and Technology of North-Rhine Westphalia.

Andreas Wieder is Secretary at the International Department of the Coal Mining Union, Bochum. (Industriegewerkschaft Bergbau und Energie).

ACKNOWLEDGEMENTS

We owe several debts of gratitude to those who made possible the 1993 conference and the publication of its papers in their current form. None of this would have been possible without the sympathetic guidance of Ray Cunningham, Projects Director of the Anglo-German Foundation for the Study of Industrial Society. The recruitment of German speakers to the conference was largely at the behest of Wolfgang Steingräber. The organisation of the conference and its budget was executed with charm and efficiency by Grace Roberts of Sheffield Hallam University's Business Services. Sue Bragg coped admirably with a miscellany of computer disks and software to produce the final manuscript; Pamela Hibberd provided secretarial support at all stages of the project; Jonathan Grove provided additional editorial assistance and produced the maps which accompany these papers. We are particularly grateful to Salford University's Services for Export and Language for their highly professional work in providing simultaneous translations at the conference itself and subsequently very speedy translations of the complete papers in German. Finally we should like to thank all the contributors for participating in the conference and refining their papers for publication. The editorial task has continued the work of the conference in providing a practical example of Anglo-German co-operation.

Chas Critcher
Klaus Schubert
David Waddington
July 1994

EDITORS' PREFACE

By 1992 we had been involved in studying conflict and decline in British coalfields for over eight years. Our initial interests were in the Yorkshire coalfield around Sheffield but these inevitably extended to the rest of Britain. We became increasingly aware that the dilemmas facing the industry were replicated across Western Europe, especially in Belgium, Spain, France and Germany. Research by the Coalfields Community Campaign especially drew our attention to what appeared to be a stark contrast between the handling of coalfield contraction in Britain as compared to Germany.

In 1992 we had established contact with Klaus Schubert at Duisberg University and together we submitted a proposal to the Anglo German Foundation for the Study of Industrial Society to undertake a comparative analysis of coal contraction in Britain and Germany. The trustees took the view that there was value in such a comparison but its impact would be dissipated by a long-term research project. They felt something more immediate was needed, which would bring together policy-makers from each country to exchange experience. They therefore invited us to submit a proposal for a bilateral conference on coalfield regeneration. The proposal was accepted, £25,000 was allocated and the conference took place in Sheffield in June 1993.

The format of the conference involved quite short presentations. Some were prepared especially for the conference and tailored to its needs. Others were summaries of longer tabled papers. Hence the contributions of the present volume vary in length. Though we have attempted to impose some similarity of subheadings on the papers, they are otherwise generally as submitted. The one exception is that expense precluded the reproduction of the slides with which several contributors illustrated their talks. Where we have had to undertake sub-editing, we have endeavoured to retain the sense and tone of the originals.

This volume is divided into six main sections. In Part I there are two overviews. David Waddington and David Parry discuss the background to coalfield contraction in Britain, whilst Klaus Schubert and Martin Bräutigam provide the same for Germany. Both papers demonstrate the complex interweaving of economic and political factors in determining the future of the coal industry.

Part II concentrates on industrial regeneration at a regional level. Matthias Hessling explains the strategic role of Ruhrkohle in the concentration, contraction and diversification of the coal industry in the Ruhr. Hedley Salt describes the Dearne Valley partnership, an attempt initiated by local authorities to regenerate a comparatively small area of Yorkshire blighted by pit closure. Ray Hudson takes a critical and retrospective view of a previous attempt at industrial regeneration of an area in Durham badly affected by steel closures, themselves following earlier pit closures.

In Part III, the emphasis on industrial regeneration is continued, but with a look at more specific measures intended to combat the effects of pit closure. Franzis Roxlau-Hennemann and Hans Esterman explain in detail the work of the Eastern Ruhr Development Agency in reclaiming land left derelict or contaminated by the closure of pits. Wolfgang Steingräber provides a history of the Zukunftsaktion Kohlegebiete (ZAK), a campaigning coalition of interests formed to defend and advance the interests of coalfield areas in Germany, which now has an extensive network of European contacts. From Britain, David Pickering, Commercial Director of British Coal Enterprise, explains their efforts to retrain ex-miners and foster the growth of new and existing businesses in coalfield areas.

The theme of Part IV is 'fostering entrepreneurship'. Wulf Noll outlines the series of programmes initiated by the Land of North-Rhine Westphalia to cope with coal contraction, emphasising the need to decentralise and the necessarily long-term nature of restructuring the local economy. Rainer Buhr analyses innovative forms of product development. These involve more than finding new markets for old products, since new industries can also be fostered. He cites the example of recycling obsolete motor vehicles. Pat Richards from North Nottinghamshire TEC explains the principles and practices involved in using government aid to the coalfields to restimulate the local economy rather than assuming retraining the workforce to be an adequate solution.

Part V looks at environmental issues. Andreas Wieder explains attempts to cope with the quite horrendous environmental legacy of mining in the former GDR. Friedrich Wagner outlines the legal and financial measures being used to reclaim land in North-Rhine Westphalia and to control the potential effects of open-cast mining. Ewart Parkinson's account of regeneration in South Wales is highly sensitive to environmental issues of decay which have only been partly addressed to date.

Part VI concentrates on issues around education and training. Heiko Beenken details the range and complexity of the measures initiated in the Ruhr to cope with the potential redundancy of miners, including short-time working, retraining and subsidies to businesses re-employing ex-miners. Roland Schäfer focuses on education and training in the small town of Bergkamen in the Eastern Ruhr. Case studies are provided of good and bad practice, with a clear recognition that training now has to be a life-long process. Hywel Francis examines the potentially strategic role of education and training institutions in regenerating both the economy and the culture of South Wales.

In a brief afterword, Chas Critcher outlines what he takes to be the key issues emerging from these accounts and attempts to relate clear differences of approach between the two nations to their political structures and ideologies.

We believe that the conference and these papers do provide an invaluable insight into the effects of coal contraction and the quite different kinds of response by elected bodies in the two countries. Nobody, least of all our German contributors, would claim that North-Rhine Westphalia has managed to solve the complex problems of coping with pit closure and restructuring the local economy in coalfield areas. There are some frank admissions about the limitations of such schemes, the difficulties of changing the culture of employers and employees and, not least, continuing political and economic uncertainty over what little is left of the German coal industry. Nevertheless there is, for us at least, a clear message to and for ex-coalfield areas which comes across from Germany and, in a different way, from South Wales. This is that there is an alternative strategy to allowing market forces alone to decide the future of coalfields areas. We owe it to ex-miners, their families and communities to keep alive an alternative vision of the future from the sense of helplessness which otherwise pervades discussion of the prospects of ex-coalfield communities.

PART I
THE CONTEXT

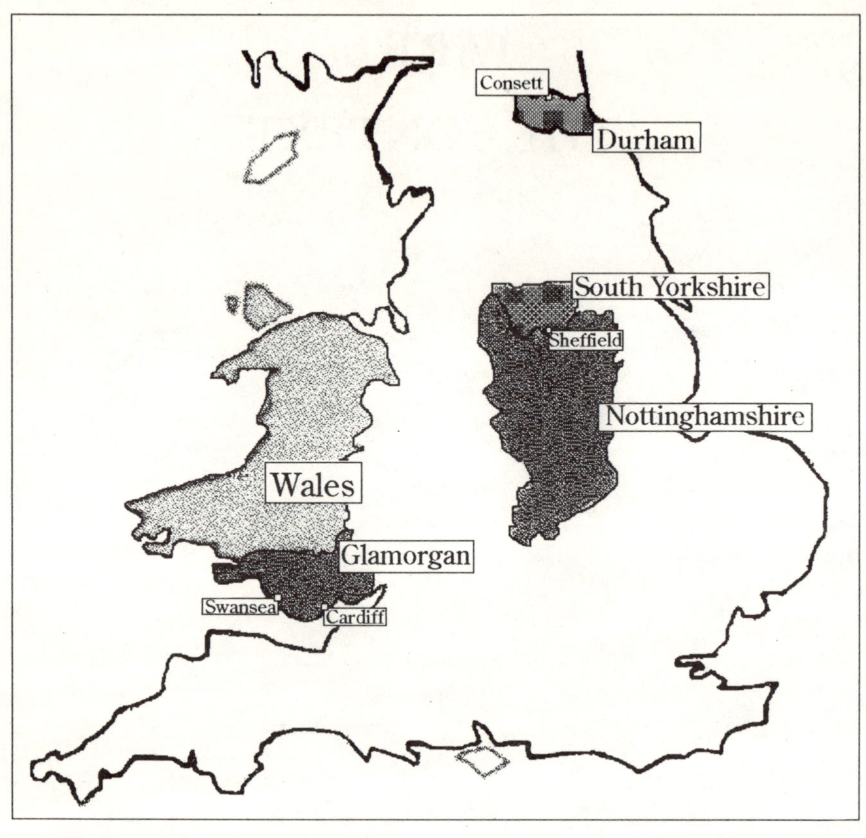

Map 1: Map of England and Wales showing ex-mining regions discussed in the text

Map 2: Map of Nordrhein-Westfalen showing towns and cities discussed in the text.

1

COAL POLICY IN BRITAIN: ECONOMIC REALITY OR POLITICAL VENDETTA?

DR DAVID WADDINGTON AND DAVID PARRY

Introduction

In its heyday of the 1920s, the South Wales coalfield boasted no fewer than 271,000 miners. As any map of the period would emphasise, scores of tiny valley communities huddled around a dense array of local mines (cf. Francis and Smith, 1981). A hotbed of socialist politics, the region's distinctive proletarian culture has been celebrated and romanticised by literary figures, like A.J. Cronin, Lewis Jones and Richard Llewellyn, and in Hollywood feature films such as *Proud Valley* and *How Green Was My Valley*. The South Wales coalfield has been a unique breeding ground for great socialist politicians, such as James Griffiths and the immortal Aneurin ('Nye') Bevan – both of whom were members of the 1945–51 Labour Government whose crowning glory was the creation of the post-war Welfare State. Not surprisingly, then, 'To the outside world the South Wales Valleys are still synonymous with coal. Yet this image, born of a century old tradition, has long ceased to correspond with economic reality' (Morgan and Price, 1992, p. 3).

By the autumn of 1992, the progressive and accelerating demise of the South Wales coalfield was virtually complete: a mere three working pits remained, two of which were scheduled to close by the following Christmas. In the words of one commentator, history was 'marching backwards' in South Wales:

In the mid-19th century, the valleys were green and rural and the Rhondda was described by a traveller as the 'Alps of Glamorgan'. Demand for King Coal led to frenzied immigration and a population explosion...

Between the 1870s and 1920s, the numbers of miners leapt from a few thousand to more than a quarter of a million. A few years after the second world war, that number had halved to just over 100,000. By 1984, the year of the strike, there were only 21,500 miners left in South Wales. Defeat and two Conservative general election victories sealed their fate. (Baxter, 1992, p. 12)

The year-long miners' strike of March 1984 to March 1985 was – as even its fiercest detractors might concede – a valiant attempt by mining families and their supporters to oppose the closure of the industry's 169 mines and safeguard the future of those communities who virtually depended upon them for their existence (Samuel *et al.*, 1986). The defeat of the principal mining union, the National Union of Mineworkers (NUM), paved the way for an unrelenting pit closure programme which bore all the hallmarks of an attempt by the Conservative Government to exact revenge on the miners for two humiliating strike defeats in the 1970s which contributed to the loss of a general election (Crick, 1985; Wilsher *et al.*, 1985). The recent memoirs of former Tory cabinet ministers confirm that the running down of the mining industry was undertaken with the objective of forever nullifying the perceived political threat posed by the NUM, epitomised by its passionately outspoken national president, Mr Arthur Scargill (Lawson, 1992; Parkinson, 1992; Walker, 1991).

Between March 1985 and September 1992, 119 British collieries were closed, including what had previously remained of the Kent coalfield. The alarming and bewildering changes occurring in the South Wales Valleys were also being experienced throughout the wider UK, not least in the other so-called peripheral mining areas of Scotland and the north-east of England. Then, on 13 October 1992, the Secretary for Trade and Industry, Michael Heseltine, unveiled the Government's plans to close 27 more collieries and 'mothball' another four; but nothing could have prepared the Government for the nationwide anger and indignation aroused by this announcement. Such outrage was soon manifested in massively supported street demonstrations (occurring in even the surest of Conservative electoral strongholds) and a backbench 'rebellion' by Tory MPs.

Clearly unsettled by the nature of public reaction, the Government authorised a pair of Parliamentary Select Committee Inquiries, on employment policy, and energy policy and the market for coal (HM Government, 1993a; 1993b). However, it is now apparent that this was merely a stalling device. Despite the unflagging public support volunteered by campaigning groups like Women Against Pit Closures (cf. Milne, 1993), and the recommendations of the Commons and Industry Select Committee, which produced a viable basis for preserving most of the threatened collieries, a government White Paper of 25 March 1993 temporarily reprieved only 12, and mothballed another six, of the 31 collieries on its so-called hit list (Department of Trade and Industry, 1993). In granting this concession, the Government managed both to appease its backbench rebels and defuse public opinion.

Thereafter, British pits closed with such accelerating regularity that, by December 1993, only four of the 31 named collieries remained operative. Among those affected were Markham, until then the only surviving colliery in Derbyshire, and Betws and Taff Merthyr, two of the three previously remaining pits in South Wales. It was not long afterwards – in April 1994 – that the region's sole surviving mine, Tower colliery, was also claimed by the closure programme.

At one time it looked as if Tower might survive in the face of seemingly impossible odds. After intense campaigning, including a 280-mile march by a delegation of local miners to deliver a letter to Michael Heseltine and an underground sit-in by the Labour MP for Cynon Valley, Ann Clwyd, it appeared that the miners had won a famous reprieve for their colliery. Local NUM lodge members had voted to resist closure by putting the future of the mine into British Coal's Modified Colliery Review Procedure (MCRP), an adjudication process which could keep the pit operative for up to nine months. Almost immediately, British Coal withdrew Tower from the MCRP and declared that production and development would be continued for another year, at which point the pit would be offered up as part of any privatisation package.

Subsequently, however, when British Coal let it be known that the conditions for keeping the colliery open would be reductions in basic wages and bonus payments and the possible loss of any future redundancy entitlements, the members voted again – this time to accept the closure of the mine. The employer's tactics, which were widely interpreted as an attempt to sabotage the review process while coercing the miners into accepting the closure of their colliery, provoked an angry reaction by Ms Clwyd:

British Coal, aided by this government, have waged psychological warfare on the Tower miners, changing the goalposts at every single opportunity to undermine their resolve. I am very angry but don't blame them for reaching the decision they have. (*The Guardian* 20.4.94)

What has been happening in South Wales represents a microcosm of what has taken place throughout the remaining UK coalfields. In this chapter, we begin by tracing the history of the British coal industry and government coal policy from the mid-nineteenth century to the Conservative's election victory in 1979, emphasising the relationship between key contemporary developments and the various historical attitudes and animosities which shaped them. Following this, we explore the political agenda underlying the coal industry policies pursued by successive Conservative administrations, under Margaret Thatcher and John Major, since 1979. The third and fourth sections consider the manner in which pit closures have been implemented by British Coal and the damaging impact they have had on mining communities. A further section critically evaluates the various ameliorative and regenerative strategies which have been designed to offset the negative consequences of pit closures on individuals, families and mining areas generally. Finally, we set an agenda for the remainder of this book by summarising some of the alternative strategies for social and economic rejuvenation favoured by Britain's EC coalmining partners, notably Germany.

King Coal's troubled reign: 1830–1979

It is only as recently as 1838 that one-quarter of all British industry's generative power was provided by hydro energy; but even by that time a spectacular change was already underway. Between 1830 and 1880, the annual output of coal grew by over 600 per cent, rising from 23 million to 147 million tonnes. This reflected its new-found status as Britain's primary source of industrial energy:

Within the space of half a century, coal was truly king and the men who were winning it had become politically and socially one of the most important and catalystic elements of the British working class. (Hall, 1981, p. 20)

The growing domination of steam power in industry ensured the continuing expansion of coal. By the peak year of 1913, coal production stood at 287,430 million tonnes per annum; there were 3,024 pits and 1,127,900 miners (Allen, 1981, p. 13). However, the highly fragmentary way in which the industry was owned and controlled, involving fierce competition between scores of small owners, led to a succession of unplanned and uncoordinated rounds of pit closures. By World War II, the industry had lost a half of its mines and one-third of its miners, owing to the destructive effect of cyclical trade depressions:

The coal industry was hit particularly hard in the inter-war years. Almost one quarter of insured mineworkers were wholly unemployed in 1934. This was an average figure. In Wales the miners' unemployment rate was 35 per cent. Some miners barely entered a pit yard there in twenty years. (Allen, 1981, p. 13)

The Coal Industry Nationalisation Act 1946 had the effect of transferring all the assets, rights and liabilities of the coal industry to the National Coal Board (NCB) and provided compensation for all private coalowners. The conversion to nationalisation in 1947 was celebrated as a development capable of delivering a planned energy policy and enhanced security for Britain's 980 pits and the 718,400 employees who mined them.

Things did not turn out quite as the miners had anticipated. Over the next ten years the NCB gradually eliminated many of its smaller, 'less viable' pits. By 1957, it had closed 158 mines. In the same year the consumption of coal fell by five million tonnes, but even this sudden slackening of demand gave no indication that the mining industry was about to enter a ten-year period of decline which would see coal becoming progressively displaced by oil as the nation's primary fuel.

Between 1957 and 1969, 505 collieries were closed with a loss of 378,000 jobs. In 1967 alone, 34,000 miners were affected in some way by pit closures. There were 12,000 redundancies in that year, six times higher than in any previous year. The worst affected areas were the 'peripheral' coalfields of Scotland, Durham, Northumberland and South Wales. This decline was partly the result of increased mechanisation which had yielded tremendous improvements in output. However, the principal cause lay elsewhere:

The proportion of Britain's fuel needs met by coal declined dramatically. After the war, nine-tenths of the country's energy came from coal: by the early 1970s, little more than a third did. The crown had been seized by oil. (Hall, 1981, p. 30)

The manner in which pit closures were implemented has a now-familiar ring. Hall talks, for example, of the 'great bitterness' caused by the off-hand and arbitrary way in which closure decisions were made and executed – invariably without any negotiation. Great dismay was expressed by miners and their trade union officials as, time and time again, pits were closed and valuable reserves were locked away for an eternity: 'It appeared scandalous to lose pits today that would be economically and financially sound, because oil enjoyed a temporary price advantage' (Hall, 1981, p. 113).

Throughout this period, miners and their union representatives had been conditioned by to see the closures as inevitable, an immutable fact of life:

The manner in which pits were closed encouraged an attitude of fatalism. Almost anything could happen and nothing could be done about it. The general explanation encouraged this attitude. Everything was all so self-evident. It was all happening through the cold, rational, impersonal, uncontrollable operation of the price mechanism. Oil was cheaper than coal; the demand for coal, therefore, must fall. (Allen, 1981, p. 60)

Starting from 1967, however, left-wing NUM officials and rank-and-file activists, notably in Yorkshire and South Wales, had begun a process of political education and opposition to pit closures (cf. Francis and Smith, 1981, pp. 425–85; Taylor, 1984, pp. 163–211). The centralisation of wage bargaining structures fostered a greater sense of unity among the NUM membership whose new militancy was reflected in two national strikes of 1972 and 1974 (Winterton and Winterton, 1993).

In the former case, the closure of the huge Saltley cokeworks in Birmingham by a mass picket orchestrated by the future NUM president, Arthur Scargill, was the centrepiece of a successful 'flying picket' strategy which secured a favourable strike settlement for the miners. During the 1974 dispute, the Prime Minister called a general election based on the slogan 'Who runs the country?', which the Conservatives lost, having misjudged the mood of the electorate. The election of a Labour government and the onset of a world oil crisis, which encouraged a greater reliance on coal, instilled greater opimism for the future of the industry. The National Coal Board announced plans to improve the output of coal by 42 million tonnes a year – primarily by extending the lives of collieries which might otherwise have closed, developing existing pits and sinking brand new collieries.

Meanwhile, the spectre of Saltley and the downfall of the Heath Government continued to haunt the Conservative Party for the rest of the decade (Scraton, 1985). The Tories' period of opposition was spent contemplating a possible strategy for any future showdown with the miners. A leaked document, prepared by the Conservative policy group on nationalised industries (the 'Ridley Report'), oulined a strategy for such a confrontation. The report advocated that any future Conservative cabinet should stockpile coal at the power stations; arrange contingency plans for imported coal; introduce dual oil-and-coal fired burning in power stations; reduce the social security entitlements of strikers; and establish mobile police squads to deal with flying pickets; and secure sufficient numbers of 'reliable', non-union drivers who would be prepared to cross picket lines (*The Economist* 27.5.78). The Conservatives were elected in May 1979, under the leadership of Margaret Thatcher. Five more years were to elapse before the cataclysmic confrontation that both the miners and the Tories were anticipating was finally realised.

Conservative coal policy since 1979

Writing within the first two years of Margaret Thatcher's first term of office as the Conservative Prime Minister, the British sociologist Vic Allen remarked that:

Coal is the earth's most plentiful energy resource. It has been estimated that 90 per cent of the world's fossil fuel comprises coal. Britain, in particular, is well endowed with this hydro-carbon energy source. Although we have worked coal seams intensively for more than 100 years there are reserves for at least another 100 years with the present methods of exploitation. If the methods are improved to enable thin seams to be exploited and to gain access to hitherto inaccessible seams then there should be enough coal for hundreds of years more. (Allen, 1981, pp. 1–2)

In this section we endeavour to explain why successive Conservative Governments under Mrs Thatcher (1979–91) and John Major (1991–) have chosen to sacrifice these abundant natural reserves in preference for more expensive and unstable sources of energy. There can be little doubt that the present crisis confronting the British deep coal mining industry is the direct product of over a decade of sustained Conservative anti-coal policy. The Conservative Government has deliberately prioritised nuclear power over coal, largely on the basis of political rather than strategic energy considerations:

We conclude that throughout the period, the underlying policy objective for coal was primarily to break the power of the NUM and the perceived stranglehold of coal on the Electricity Supply Industry, but also to demonstrate the failure of public ownership.... From the beginning, Government policy was to expand nuclear power, in part if not wholly as a means of achieving the political agenda on coal. (Parker and Surrey, 1992, p. 58)

This political agenda was undoubtedly in place when the Thatcher Government was first elected in 1979. However, as Parker and Surrey point out, circumstances were not immediately propitious for a direct confrontation with the NUM. Ministers were wary of the NUM's power to disrupt electricity supplies and inflict another harmful defeat on their administration. In any case, the Government was keen to further reduce its dependency on oil at a time when imported coal was relatively expensive and there were no viable alternative sources of energy to call upon.

Notwithstanding these constraints, government spokespersons indefatigably emphasised the need for greater competitiveness and efficiency in the coal industry, insisting that, before too long, it must be able to 'stand on its own two feet', independently of government grants or subsidies. The onset of recession and a corresponding decline in the demand for industrial energy coincided with an upsurge in the world supply of coal from new producers like Columbia (at a price attractive to key consumers like the British Steel Corporation), while the implementation of new technology in British mines was greatly expanding their rate of production and leading to overcapacity (Winterton and Winterton, 1993).

Information leaked to the NUM in February 1981 indicated that the NCB were set to respond to this growing crisis – and the strict financial constraints currently being imposed upon them under the Coal Industry Act 1980 – by shutting down 50 collieries. Shortly afterwards, the Board unveiled proposals for the 'accelerated closure' of 23 collieries. The NUM immediately gave notice of its intention to ballot its members on strike action. This threat, reinforced by a rash of unofficial action throughout the coalfields, forced the Government to back down. The proposed closures were withdrawn; ministers agreed to stem imports and take financial measures to offset the NCB's current operating deficit.

The Government's response amounted to a stalling tactic providing time to prepare more adequately for a confrontation with the miners. The problem of uneconomic production was unresolved; and when, in June 1983, a report by the Monopolies and Mergers Commission identified 141 out of the 198 collieries as unprofitable, a major closure programme seemed imminent. The appointment of Ian MacGregor as NCB chairperson on 1 September 1983 made it crystal clear that the mining unions were about to be confronted in the same way that steel and car workers had been challenged (and defeated) during MacGregor's preceding appointments at British Steel and British Leyland.

Within a month of MacGregor's appointment, the NUM imposed an overtime ban in response to the NCB's 'final offer' of a 5.2 per cent wage rise in return for co-operating in the closure of high-cost collieries. Then, in March 1984, as unofficial strikes – caused by managerial sanctions against the overtime ban – spread throughout South Yorkshire, the NCB stated their intention to close the Cortonwood colliery near Sheffield. Days later, the Board disclosed further plans to close 19 other collieries that year. Two days afterwards, on 8 March, the NUM National Executive lent its approval for the militant Scottish and Yorkshire areas to take strike action under Rule 41 (Winterton and Winterton, 1989).

The National Executive had confidently assumed that this decision would have the 'domino effect' of encouraging NUM members in other coalfields to sympathetically join the strike while, at the same time, obviating the need for a national membership ballot. However, the refusal of the strategically important Nottinghamshire miners to join ranks greatly impaired the NUM's capacity to win the strike. It was this factor, allied to the strike-breaking role of a nationally co-ordinated police force (Fine and Millar, 1985), and the failure of the NUM to win over public opinion in the face of a hostile mass media (Jones *et al.*, 1985), which led to the defeat of the year-long strike and enabled the Government to exact full revenge for the humiliations of 1972, 1974 and 1981 (cf. Wilsher *et al.*, 1985 for an overview of the main events).

The subsequent decision by the Nottinghamshire area to break away from the national union and form the Union of Democratic Mineworkers (UDM) left the NUM all the more weakened and demoralised. While initially appearing to favour the breakaway union, British Coal (which was renamed in March 1987) ruthlessly pressed home its advantage over the coal unions. Over the next seven years, the number of deep mines was reduced, from 169 to 65, and the industrial workforce pared down from 171,000 to 57,000. By the end of this period, British Coal was showing an overall operating profit after interest, independently of any operating subsidy. However, as Parker and Surrey (1992, p. 18) point out, there has been no attempt to 'weigh the costs of restructuring and redundancy against the wider social and unemployment costs'.

It has long been the avowed objective of Conservative ministers to return a profitable coal industry to the private sector (Winterton and Winterton, 1993). At the Tory Party Conference in October 1988, Cecil Parkinson triumphantly proclaimed that a re-elected Conservative Government would carry out the 'ultimate privatisation' of the coal industry. As he later elaborated in his memoirs,

What was ultimate about the proposed privatisation of coal was that it would mark the end of the political power of the National Union of Mineworkers and would make the coal industry what it should always have been, another important industry, no more and no less important than many others. (Parkinson, 1992, p. 280)

In pursuing this overtly political objective, successive Conservative governments have wilfully favoured the development of nuclear power in order to marginalise the coal industry.

Parker and Surrey use leaked Cabinet minutes and ministerial memoirs (cf. Lawson, 1992) as proof that, prior to 1988, the Conservative Government's commitment to the nuclear power industry was motivated less by objective economic arguments than a political objective to undermine the power of the NUM.

As the economic argument in favour of coal and against nuclear power has become progressively more compelling (for example, in terms of cost and safety), government rhetoric has focused increasingly on the greenhouse effect and the alleged unreliability of world fossil fuel prices – two problems which, they argue, are best countered by encouraging nuclear power. During the late 1980s,

Coal-fired power stations were often portrayed as the main danger to the planet even though they were responsible for less than one tenth of the total emissions of greenhouse gases worldwide. This mood was reflected in the Government's White Paper on the Environment, *Our Common Inheritance*. The only references to coal were as a source of pollution: there were no suggestions as to how this could be mitigated. There was no reference to clean coal technology, particularly of the potential of such technology to reduce emissions and BCC's internationally acknowledged expertise in this field. Coal was seen as a liability, not an asset. (Parker and Surrey, 1992, p. 23)

The recent privatisation of the Electricity Supply Industry (ESI) provides further evidence of a deliberate policy to promote nuclear power at the expense of coal. Until 1991, the public sector ESI constituted a captive market for the coal industry, accounting for some three-quarters of all British Coal's sales. With privatisation, the monopolistic Central Electricity Generating Board was divided up into two central private generating companies, PowerGen and National Power, and a state nuclear power company, Nuclear Electric. The privatisation process also created 12 Regional Electricity Companies (RECs), with an eight-year monopoly of local electricity distribution markets.

An element of competition was encouraged by allowing the RECs to seek alternative electricity sources to that provided by the main generators. In the late-1980s the EC lifted its ban on the use of gas, supplies of which had hitherto been held back for conservation purposes. Being anxious to avoid over-dependency on the two large generating companies, and in looking to give themselves a bargaining lever, the RECs have commissioned the building of some 20 gas-fired power stations. These stations have the advantage of being relatively cheap to construct and are capable of providing cleaner power than current coal-fired stations.

This latter consideration is undoubtedly important, given the tightening of EC directives on sulphur and carbon dioxide emissions. However, it is difficult to see why the RECs (not to mention the Government) should have so readily allowed this to override other important factors, such as the relatively short lifespan of gas supplies, and the fact that coal-fired power stations can produce electricity up to 30 per cent cheaper than that provided by the gas-fired replacements (Coalfields Community Campaign, 1993).

A further problem created for the coal industry by electricity privatisation is that independent power stations, including the French state-owned nuclear industry, have sought to compete in the supply of electricity. The two main generators have reacted to such threats by also participating in the 'dash for gas', and have stepped up their intake of imported coal and Venezuelan orimulsion, which the Government has agreed to let them burn despite its reputation as the world's dirtiest fuel (*The Guardian* 14.8.93).

The strong disparity in the Government's treatment of the coal and nuclear power industries clearly pervades this whole process. As part of the privatisation settlement, nuclear power was awarded a guaranteed market share, under the 'non-fossil fuel fraction' created by the Electricity Act 1989, while the imposition of a fossil fuel levy of more than 10 per cent on every home and business in England and Wales is meant to be used to subsidise the decommissioning costs of closing down old plant. Fothergill (1993, p. 170) points out that this subsidy is actually being used to finance the day-to-day operations of Nuclear Electric.

European Commission rules specify that each member state is permitted to support only one power-generating industry. It is symptomatic of the Government's prejudicial attitude toward the mining industry that it has chosen to subsidise nuclear power while Germany is supporting coal – despite the fact that British coal is produced at half the cost of German coal (McAvan, 1993), while nuclear power is easily the most expensive method of generating electricity (Coalfields Community Campaign, 1993).

By the autumn of 1992, it became apparent that the Government was set to agree a renewed five-year contract with the generators which anticipated a sale of 40 million tonnes in 1993–94, falling to 30 million tonnes in the remaining four years. This compared with an agreed supply of 65 million tonnes for 1992–93. The Government not only refused to provide any direct subsidy or other form of protection from the 'dash for gas' (such as ceasing to approve new gas power stations), but also continued to endorse the creation of coal import terminals, notably at Immingham, thus dealing another death blow to the industry.

It was this expectation of reduced sales to the generating companies which set the context for Michael Heseltine's announcement of 13 October 1992 that 31 deep coalmines were to be closed or mothballed. The widespread public consternation triggered by this announcement appeared to be shared by the European Commission who were said to be 'unhappy with a situation where the lowest-cost Community coal producer faced rapid contraction while high-cost German and Spanish production continued at a high level' (Parker and Surrey, 1992, p. 28). The Government's temporary suspension of the closures reopened the possibility of economic debate. The report by the Select Committee on Employment established a formula whereby the provision of direct subsidies both to British Coal and the power generators, the halving of opencast production and a ban on imported French electricity could ensure a viable future market for most, if not all, of the threatened mines. Now it was up to the Government.

On 25 March 1993, the Government duly published its White Paper on the future of the coal industry. At first sight it gave the impression of being 'a good old-fashioned British compromise' (Fothergill, 1993, p. 169), in that it provided for six of the 31 mines to be mothballed and a further 12 to continue their stay of execution. These 12 collieries were to be kept open for a period of 'market testing': to see whether additional custom could be obtained for the 'surplus' coal they were currently producing. The White Paper also set out a package of support measures for the coalfield communities, including a £75 million training, counselling and job search programme, and a separate amount of £75 million for the provision of sites and premises (DTI, 1993).

The contents of the White Paper were designed to mollify political opposition from Conservative MPs, the media and public opinion. Few informed observers gave the 'market tested' or mothballed collieries any realistic chance of long-term survival, given that government policies were already producing a shrinking market for coal (Fothergill, 1993). By mid-December 1993, no fewer than 27 of the 31 mines had been closed down – even though British coal was being mined much more cheaply than imported foreign coal (*The Guardian* 20.10.93) and the economic and environmental case against nuclear power was gaining support. A political vendetta was being pursued.

Managing pit closure

The sudden and secretive manner in which the British Government announced the acceleration of its pit closure programme in October 1992 typified its total disregard for any form of consultation and debate, and its lack of consideration for the social consequences of its policy. As McAvan (1993, p. 174) points out, press speculation concerning possible pit closures had been consistently denied by ministers – despite leaked correspondence to the contrary – since the general election of the previous April. At no time did the Government offer or propose to engage in consultation with trade unions or local authorities representing mining areas liable to be affected by pit closures.

Currently there is no formal consultation between British Coal management and national NUM officials. The latter are boycotting such talks due to the employer's refusal to allow the NUM to represent its members in the UDM-dominated Nottinghamshire and Midlands areas. In theory at least, there is no such impediment at local level where miners are entitled either to accept a proposed pit closure or request that the future of the colliery be determined via the Modified Colliery Review Procedure (MCRP), whereby the employer and trade unions set out the cases for and against pit closure.

No colliery has ever remained open as a result of the MCRP because British Coal have never accepted the outcome as binding. Even when it has been recommended that the mine continue working, as in the case of the Bates (North East) and Cadeby (South Yorkshire) collieries in 1986, the employer has tended to disregard the decision. British Coal steadfastly refuses to accept that it has any moral obligation to consider the social consequences of closing a mine for a single community or entire mining area (Beynon *et al.*, 1991, pp. 87–90).

Since the autumn of 1991, we have been visiting a number of Yorkshire pits where miners have been voting whether or not to accept pit closures. Their decisions have typically been framed by a number of familiar considerations. First of all, miners have expressed anger and disenchantment at their post-strike treatment by management, who have constantly employed the threat of pit closure to extract higher productivity. Secondly, employees have grown intolerant of constant job insecurity which forbids any planning for the future. A third factor has been the speed with which miners are required to arrive at their decision – usually whether to vote to put the colliery into the MCRP or accept the enhanced redundancy terms temporarily on offer – without being given adequate time to contemplate not just their own futures but those of their dependants and the wider community also (Waddington *et al.*, 1994, pp. 144–5).

The potential for collective action had all but disappeared: NUM branch officials were aware that employment legislation had effectively outlawed 'secondary picketing' and rendered people like themselves legally culpable for the deployment of pickets. Besides which, each branch was conscious that it would be suicidal to engage in militant action as British Coal would use the men's 'unfavourable attitude' as a spurious basis for closing the mine.

There is plentiful evidence that, since the Heseltine announcement, British Coal has consistently resorted to threats and bribery to encourage mineworkers to accept enhanced redundancy packages rather than taking their pit into the MCRP. An example was Calverton colliery in Nottinghamshire in November 1993, where miners voted by a 2:1 majority to enter the MCRP. British Coal instantly responded by further enhancing the redundancy package by £7,000 per employee – but only on condition that the men agree to an immediate closure. Three days later, on 10 November, a reconvened branch meeting was held at which members reluctantly accepted closure (Coalfields Community Campaign, 1993). Such practices have led NUM officials to accuse British Coal of 'blackmail' in their dealings with the workforce, and to castigate the Government for wantonly disregarding the social consequences of its policies.

The social impact of pit closure

We emphasised in a previous publication (Dicks *et al.*, 1993) that apocalyptic media representations prophesying the 'sudden demise' of mining communities were guilty of obscuring a less dramatic, though arguably more disturbing, reality for coal-mining areas affected by the Heseltine announcement. As we pointed out, a mine closure in the 1990s merely puts the final seal on what might already have been a decade or more of 'quiet disintegration'.

This point is best illustrated with regard to the South Yorkshire mining village of Grimethorpe whose pit was closed in 1993. It had been less than three months after the miners' strike of 1984–85 that the journalist Sam Miller visited the community and made the following observation:

Grimethorpe is a village which suffers from urban decay. It is a small place surrounded by fields and farms and other villages, yet it suffers from vandalism and glue-sniffing – activities associated more often with blighted inner-cities. Glue-sniffers, all children from the village, were discovered at the condemned remains of the Grimethorpe Empire. This discovery was made only six weeks ago and has clearly stunned the village. (Miller, 1985, p. 16)

Significantly, Miller discovered that the pit had not taken on any new workers for two years and that there were strong rumours of the NCB's intention to close the colliery within the next three years. As many as one-fifth of the workforce were likely to take redundancy in the next 12 months. All of this suggested a possible correlation between vandalism and solvent abuse and the absence of job prospects for local youth.

Pit closure programmes – which tend to be preceded or accompanied by the corresponding contraction of other staple industries like engineering, steel or transport – can bequeath to an entire former industrial heartland an unwanted legacy of high unemployment, a low-wage economy characterised by a low female activity rate, environmental decay, and an unenviable rise in crime and vandalism resulting from shattered hopes and disaffection among the young (Morgan, 1993; Morgan and Price, 1992; Morris and Wilkinson, 1993). Single communities may find themselves blighted for years by the impact of pit closure. This is graphically illustrated by what has happened since the closure of Langwith colliery in Derbyshire in 1978:

While the pit was active the NCB had assumed responsibility for the recreation ground and the cricket pitch; had provided an area of allotments for miners; had kept the Miners' Welfare in good condition and carried out a number of maintenance and tidying operations around the colliery environs. All of these services disappeared with consequent costs in social and environmental terms. (Derbyshire County Council, 1985, p. 27)

The former colliery site remained a derelict and abandoned eyesore for years afterwards while, in the community itself, vandalism increased – a development which the local council attributed to receding employment prospects for young people.

It is therefore apparent that the impact of pit closure cannot be properly understood without adequate reference to prior and future processes of industrial and social decay. The most immediate impact of pit closure is that hundreds of people are likely to be thrown out of work. Various surveys, including one of our own, indicate that it is typical for between 25 per cent and 50 per cent of the miners displaced by any instance of pit closure to still be unemployed a year to 18 months later (Dicks *et al.*, 1983; Guy, 1994; Wass, 1989; Witt, 1990). These studies further suggest that new jobs undertaken outside the industry (most commonly in lower-skilled employment like security work, gardening, construction work, taxi driving or delivery work) invariably involve a deterioration in pay and conditions, tend to be temporary and insecure, and frequently involve an increased travel-to-work distance (Little and Reynolds, 1991).

Witt's (1990) survey of Yorkshire miners made redundant in 1988 showed that one-third subsequently returned to the industry as contractors. Obviously, the potential for miners to be absorbed by other pits as transferees or to re-enter the mines as contractors is gradually disappearing with the continuing contraction of the industry.

The knock-on or 'multiplier effect' of pit closure will also ensure that jobs elsewhere in the economy – in the mining supply industry, in rail and power, and in local shops, schools and the service industries – will disappear as demand dries up and purchasing power is lost (Fothergill and Witt, 1990; Glyn, 1989). Finally there will be a reduction of available jobs for school leavers.

The social psychological impact of pit closure will vary between individuals, families and even whole communities. Our current study of four South Yorkshire mining communities variously affected by industrial contraction, funded by the Economic and Social Research Council, provides a useful generalisation of the social consequences of pit closure. For example, while most of the men we talked to regretted the loss of the special cameraderie with workmates that uniquely belongs to mining, the majority were undoubtedly glad to have escaped from what they saw as an increasingly oppressive and exploitative authority.

We soon became aware of a longstanding legacy of anger and bitterness at the way that management had treated them. These sentiments were compounded by the virtually unanimous belief that political malice had been deployed to create an environment favourable to employers. Those men lucky enough to have secured another job were finding it difficult to come to terms with the loss of their occupational identity and the unique skills they had gradually acquired and honed while working underground. By contrast, others were suffering the debilitating anxiety that came from wondering when, or possibly if, they might ever work again.

The women we interviewed were finding it extremely difficult to cope with conditions of increased family poverty while simultaneously trying to help their male partners to deal with the stress and demoralisation that comes from being unemployed. The almost uninterrupted presence of a man around the house was universally viewed as disruptive, intrusive and threatening to the woman's independence. With the disappearance of mining and other industrial jobs, families were becoming increasingly dependent on women's wages – often earned in part-time, non-unionised jobs with poor conditions. In one of the communities we visited, a quarter of all former mining households comprised a woman working part-time, an unemployed husband and one or more dependent children. In addition to finding it difficult to adjust to their loss of earning and spending power, many former miners do not find it easy to relinquish their traditional 'breadwinner' status within the family.

The strain experienced by both parents can have an impact on their children. Adults often told us of their feelings of guilt and inadequacy at not being able to afford snacks, social outings, holidays or leisure-wear for their children. Paradoxically, mothers and fathers admitted to feelings of resentment caused by their children's apparent incapacity to understand the implications of unemployment for the disposal of family income. A scuffed pair of shoes or whether or not someone was entitled to a new pair of trainers could easily become the pretext for a blazing family row.

The loss of mining work can induce social divisions in what was formerly a tightly-knit and cohesive society. Relations between former colleagues can become rivalrous with each man trying to show how he had coped better than his erstwhile workmates since redundancy. Others, especially the long-term unemployed, felt lonely and isolated – physically and psychologically detached from village affairs. The disappearance of long-established rituals, like the annual village gala, epitomised what many regarded as a loss of community spirit.

Nothing in ex-mining areas symbolises the fragmentation of community more than the recent proliferation of petty crime, drug abuse, vandalism and 'joyriding', especially among local youth. Police, teachers and social workers in our study were unanimous in attributing these tendencies to the growing sense of hopelessness and despair among a generation who consider themselves abandoned by society.

These were, as we discovered, a proud, resourceful people who admitted to harbouring fear and pessimism for the future. Even those forced by the opportunity of employment to leave the village behind did so with obvious reluctance.

The closure of a pit devastates the local economy, leaves many ex-miners with insecure or low-paid employment, disrupts family life and damages the quality of community life. These are some of the challenges to those measures supposedly designed to mitigate the effects of colliery closure.

Ameliorative or regenerative measures

Writing at the peak of the pit closure programme of the late 1960s, Dennis *et al.* (1969, p. 11) grimly observed that, insofar as government policy towards pit closures was concerned, 'There is no overall planning of new industries or of training or of education for leisure; there is no more than marginal provision for economic security.' These authors expressed the view that 'this community without the mine and mineworkers is in danger of becoming merely an aggregate of socially isolated and culturally condemned human beings'.

A quarter of a century on, the position has improved, albeit slightly. Specific agencies and related financial mechanisms have been instigated whose functions are to prepare, retrain and relocate former coal industry employees, to foster entrepreneurship in coal-mining areas, and to revitalise local economies by creating or attracting new industries. These measures are discussed in detail by the British contributors to this volume.

Absent from this account are coalmining unions who play little part in helping their members to deal with the consequences of pit closure. This reflects a particular political stance: the largest union, the NUM, 'sees its major task in protecting jobs rather than in cushioning the effect of the restructuring exercise. It has therefore had a peripheral role in the activities concerning coalfield regeneration' (Edwards, 1991, p. 59).

British Coal has played an active though controversial role in assisting those of its employees affected by pit closure. The corporation's redundancy terms, in addition to incorporating any statutory requirements, also provide three weeks' pay (based on a ceiling of £300 per week) up to a maximum of 30 years' service. Lump sums of between £2,500 for employees under 21, and £10,000 for those aged 30 or over bring the maximum payment up to £37,000. British Coal has used both the threat of withdrawing its enhanced redundancy terms, or the promise of increasing them, as ways of inducing its employees to accept pit closure.

Employees transferring to another British Coal colliery may be entitled to a number of allowances. These include assistance with the costs of visiting the new colliery, removal expenses and help with any rent increase. In addition, transferees are granted a lump sum payment of £1,100 and three subsequent payments after six months, one year and two years, the sizes of which vary according to the distance of the new location from the former mine.

In evaluating the redundancy scheme, Edwards (1991, p. 61) points out that while it has certainly 'softened the blow' of redundancy for older workers, the fall in the average age and length of service of mineworkers means that fewer ex-employees will qualify for reasonable lump sum redundancy payments. She adds that such payments do little to assist workers not employed by British Coal who are nonetheless indirectly affected by pit closure. Nor do they contribute to economic regeneration unless re-invested in a small business. Similarly, while the transfer and mobility allowances have facilitated the transfer of 30 per cent of otherwise redundant mineworkers, the accelerating rate of pit closures means that many of these employees will face yet another experience of redundancy, having only just re-established themselves at new pits.

The employer-sponsored British Coal Enterprise, discussed by David Pickering its Chief Executive, was set up in 1985 to encourage job creation in coal-mining areas. It has since been involved in three main areas of activity. First, in trying to encourage the inception and growth of small businesses by providing low interest rates (subject to the close scrutiny of any proposed venture) of up to £5,000 per job created or 25 per cent of the initial funding required; second, by supporting Local Enterprise Agencies, non-profit making partnerships (e.g. involving BCE, banks, local government and firms in the area) which offer practical advice to people who may be considering starting up some sort of enterprise in the area; and third, by creating and supporting managed workshops and small business units which are usually housed in former British Coal premises, and are sometimes managed by BCE employees.

Further to these provisions is BCE's Job and Career Change Scheme (JACCS) which is designed to assist redundant mineworkers in their search for work or, where possible, retrain them for new employment. Once a pit closes, BCE set up one of their portable 'Job Shop' counselling and outplacement services in the pit yard. All ex-employees who so wish are then counselled, shown how to complete a curriculum vitae, and have their skills analysed. Where possible, these skills are matched with any local vacancies determined on the basis of a survey of local employers. Retraining is only made available where it is deemed necessary to improve an individual's chances of securing employment, and usually lasts for a period of 20 weeks. Ex-employees must demonstrate that they have a realistic chance of obtaining a job or being able to enter further or higher education as a condition of being accepted for training support.

BCE has made some impressive claims about its ability to generate jobs and find alternative employment for ex-miners. In its Annual Report for 1991/2, for example, BCE claimed to have resettled 25,000 former British Coal employees, a success rate of 86 per cent. The report lays claim to having created an overall figure of 76,000 job opportunities once the benefits of its managed workspace and business funding schemes are built in to the calculation.

As Edwards (1991, p. 59) points out, it is difficult to evaluate the validity of what appears to be such a phenomenal achievement.

However, the signs are that this apparent success has been greatly over-stated. There is evidence from studies undertaken in South Wales (Rees and Thomas, 1989; Trotman and Lewis, 1990) and Yorkshire (Turner, 1992) that the majority of the small businesses created or supported by BCE are not actually 'new' and tend not to be run by former miners. It is not readily apparent how far such businesses have actually made a net contribution to local employment, or have merely replaced or displaced existing firms. As Edwards (op. cit., p. 59) explains, 'Displacement is especially problematic in the service sector where new firms subsidised by cheap loans, accommodation or government grants undercut and put their competitors out of business.'

The revelation that the opportunity to run a small business is not being taken up by ex-miners is not due, as has commonly been supposed, to the lack of an 'entrepreneurial attitude' on their part. Rees and Thomas's (1989) study of redundant South Wales miners indicated that, while many respondents had contemplated running their own business, they were deterred from doing so by the harsh and uncertain nature of the local coalfield economy. Certainly there are grounds to support the argument that it would be fatal to transform coalfield areas into a haven of small businesses. Barnsley and Doncaster TEC (1992, p. 5) point out that:

During the 1980s a very high proportion of people started up new businesses in the area (between 1981–90 VAT registrations grew by 71 per cent in Barnsley and 77 per cent in Doncaster against a national average of 55 per cent), however a substantial proportion subsequently failed...57 per cent in Barnsley and 90 per cent in Doncaster were deregistered.

Any successes claimed by BCE's JACCS scheme have to be measured against the statistic that some 50 per cent of former employees do not take advantage of the service, sometimes because they are able to secure alternative employment by themselves, but also because of feelings of bitterness and distrust towards their former employer. Only one in ten mineworkers receive training under the JACCS scheme. One of the many criticisms levelled at the scheme by Trotman and Lewis (1990) is that 20 weeks is an inadequate period in which to train ex-miners for anything other than a relatively low-skilled job.

A study of redundant Yorkshire miners undertaken by Witt accuses BCE of failing to highlight the reality that former mineworkers invari-ably suffer long periods of anxiety prior to securing employment, and that, even when jobs are eventually obtained, they often tend to be tem-porary, low-paid, and unrewarding in character. Respondents talked of a tendency for BCE to offload their ex-employees into the nearest avail-able job, no matter how unsuitable. Witt was therefore led to the conclusion that:

The objective of British Coal Enterprise in relation to former miners would appear to be the transfer of men into any type of economic activity as soon as possible in order to demonstrate the limited impact of colliery closures upon mining areas. (Witt, 1990, p. 47)

The main government-funded training provision available to redundant mineworkers is the Employment Training Programme (ET), a two-year scheme based on elements of teaching and placement in industry. The corresponding Youth Training Scheme (YTS) offers similar support for school-leavers. Neither course leads to a nationally recognised qualification and both have been subjected to cutbacks in government funding. Additional problems associated with the ET scheme are that redundant mineworkers do not qualify for a place on them until they have been unemployed for six months; and, should the course extend to more than 21 hours per week, trainees may lose their entitlement to unemployment and social security benefits. Given these difficulties, very few miners receive training via this scheme.

These programmes are now administered by a network of 82 Training and Enterprise Councils (TECs) covering the whole of England and Wales. The chapter by Pat Richards describes how one TEC has responded to the prospect of large-scale pit closure. According to the Employment Service, TECS are 'business-led and community based' agencies which have executive responsibility for determining and responding to local training needs:

As well as managing on behalf of government a range of training, education and enterprise programmes, each TEC has the wider responsibility to act in a catalytic and co-ordinating role to ensure that training and enterprise activities are relevant to individual needs and local services. (Employment Service, 1992, p. 2)

It is possible that the £75 million aid package to be administered via the TECs will be used to enhance current training provision for redundant mineworkers. For the moment, the inherent problems of the ET scheme mean that very few mineworkers receive direct training and that courses run specifically for their benefit tend to be under-subscribed (Edwards, 1991, p. 79).

Local authorities are the only formal agency involved in the regeneration of coalfield areas who can justifiably claim to have a popular mandate for their activities. Hedley Salt's chapter describes how a consortium of local authorities set up the Dearne Valley Partnership in an area of Yorkshire affected by pit closures. Most local authorities have Economic Development Units which carry out research and provide expertise, funding and advice for a variety of regeneration activities, such as providing support for enterprise agencies, developing business parks, promoting tourism, running Job Shops and maintaining state education and training institutions.

The present Government is resolutely antithetical to the role and influence of local government. The acceleration of the pit closure programme has coincided with the imposition of strict constraints on local government spending. Those authorities which exceed these spending limits will find themselves in receipt of diminished funding for the following year. This makes it virtually impossible for hard-pressed local authorities to apply for EC aid since it must be matched by equal funding from their own budgets. This is an especially cruel dilemma, all the more so because it befalls the local authorities' social services departments to come to the aid of communities affected by central government's rationalisation of the coal industry.

According to Beynon *et al.* (1991, p. xix), the 1980s heralded the emergence of different forms of community politics from the institutional processes which prevailed in the 1960s, when the crisis confronting coalfield areas was managed by committees and decision-making bodies which included local councillors and trade union representatives. As central government policy remorselessly excluded these traditional forms of community representation, local politics took on new forms which Beynon *et al.* have helpfully dichotomised as either progressive or regressive in character.

In their more progressive form, the new politics were evident in: local campaigns against nuclear energy or nuclear waste; opposition to opencast mining; drop-in centres for the unemployed; women's support groups; and co-operative ventures, sometimes supported by local authorities. In the meantime, a more regressive strain of local politics emerged, epitomised by the proliferation of local development agencies, 'all concerned with attracting a slice of the cake, but uninterested in, or unable to affect its size, and downplaying other broader questions such as the quality of jobs and the general conditions of employment. Any crumbs are seen as better than none' (Beynon *et al.*, 1991, p.xix).

The chapter by Hudson and Sadler analyses in detail the claims made by development agencies to have successfully regenerated the Consett area of Durham following coal and steel closures. While enterprise and development agencies acting on behalf of heavily de-industrialised regions like the north-east of England and South Wales have attracted huge investments from Europe and Japan, the new industries are locally unaccountable and tend to involve poorly paid, low-skilled, routinised jobs for which there is no or only limited training. Such industries cannot adequately compensate (quantitatively or qualitatively) for the jobs which have steadily disappeared from the indigenous industries like mining and steel.

It is imperative, however, that the achievements and lessons of regional policy are acknowledged alongside their shortcomings (Morgan, 1993). The so-called Valleys Initiative conceived of by the former Secretary of State for Wales, Peter Walker, is discussed in the chapters by Ewart Parkinson and Hywel Francis. Originally designed as a three-year programme, the Initiative was further extended into a five-year programme by Walker's successor at the Welsh Office, David Hunt:

The key aim of this initiative was to enhance 'the prosperity of the valleys of South Wales and the well being of the people who live in them'. More precisely, the aims were to reduce unemployment by up to 30,000, improve education and training provision, remove environmental degradation, enhance housing and transport facilities and raise the quality of social and community life. Another aim was to promote partnerships between central and local government, public and private sectors, and employers and unions – in short between all the parties engaged in social and economic renewal. (Morgan, 1993, p. 178)

Morgan fully recognises the achievements made by the initiative thus far. These include the establishment of the largest derelict land clearance programme in Western Europe, repairs and improvements to some 7,000 homes, the acqusition of over 2.5 million square feet of new industrial floorspace, and the creation or safeguarding of 24,000 jobs based on £700 million of private sector investment. However, the same author has also emphasised that, 'While the aims and the values of the Valleys Programme were laudable enough, the initiative was essentially a marketing exercise in the sense that it packaged together a host of policies and projects that were for the most part already in existence' (Morgan and Price, 1992, p.11). There was academic consensus around the notion that the initiative was 'necessary, but in no way sufficient'.

The initiative's key aim of reducing unemployment by as much as 30,000 was destroyed by the impact of the recession. However, perhaps the main impediment to the achievement of objectives was the debilitating effect of a whole array of central government policies which have stymied advances made at regional level. Prominent among these are: a decline in regional expenditure of 50 per cent between 1988 and 1991; the financial weakness of local authorities who have been squeezed remorselessly due to cutbacks in local government spending; and cuts in government financial support for the TECs. As Morgan and Price (1992,

p. 12) point out, 'Without a more supportive national economic framework it will be difficult if not impossible to achieve anything other than little victories at the regional level'.

An alternative approach

There is a widespread and justifiably held belief among UK mining trade unionists that the implementation of pit closures and processes of industrial reconversion are handled in a far more enlightened and compassionate way by Britain's west European partners, particularly the Germans. Speaking from a community within the north-east of England whose colliery (Murton) had closed barely a year before, one former mineworker exhorted the journalist, John Pilger, to:

Look at what's happening in Germany.... Murton is twinned with Baersweiller. When the pit closes there the men will have had five years' notice, retraining and guarantees of new jobs. The town has been given grants for new industry. We've had none of that. (*The Guardian* 30.1.93)

Most west European deep coal-mining countries are experiencing the rapid contraction of their industries, with Belgium having recently closed its last mine and Portugal and France scheduled to do so within the foreseeable future:

But what distinguishes the pit closure programmes of our European partners from those of our own Government, quite apart from the strong economic argument for maintaining a UK coal industry, is that they involve long-term planning in terms of both energy and regional policy, as well as extensive consultations with local actors, including trade unions, employers and local authorities, prior to decisions being taken. (McAvan,1993, p. 172)

In France, where they are aiming to abolish deep coal mining by 2005, the state-owned Charbonnages de France is committed to finding alternative jobs for each miner aged under 45, as well as providing assistance with retraining and relocation. Miners at the last surviving Portuguese pit were being retrained during working hours 18 months before the mine was due to close. As McAvan emphasises, the cost to the employer of job-search and reskilling approaches of this type compares favourably with the once-and-for-all pay-off system preferred by British Coal.

In Belgium, where the last mine closed in the autumn of 1992, a ten-year reconversion plan (the 1987 'Future Contract') was implemented to ensure that adequate training and regeneration schemes were brought in to assist mineworkers in the Limburg region while pits were being carefully phased out. This project was funded by a financial infusion of 100 million Belgian francs (280 million dollars) – roughly what it would have cost the Government to subsidise the continued operation of the mines over the same period. A 'partnership' structure comprising employers, trade unions, representatives of the national, provincial and local Flemish governments, and members of relevant EC institutions was created to oversee the plan. At one point during its implementation, the pit closure programme was temporarily halted because the region's unemployment had risen above the national average.

An emphasis on joint planning and consultation has also characterised the phased closure programmes of Spain and Germany, the other primary western European deep coal-mining nations. In Spain, where mining is rendered difficult by adverse geological conditions and poor mechanisation, coal is heavily subsidised to the tune of 26.7 dollars per tonne. The high cost of Spanish coal, which averages 130 dollars per tonne, makes its coal industry particularly vulnerable in the fierce competition inherent in the EC's Single Market legislation. The Government has responded to this impending crisis by instigating a special retraining and investment programme in the Asturias region. In the meantime, a national energy plan was agreed in 1991 whereby state subsidies are being used to ensure a medium-term future for around two-thirds of the industry.

Finally, in Germany, currently the EC's largest coal industry, a joint agreement for the future of deep coal-mining to the year 2005 has been achieved between the Government, coal producers and trade unions. This 'Coal Concept 2005' acknowledges that production will decrease from its current level of 66 million tonnes to 50 million tonnes in 2005. There will be corresponding reductions in the number of working mines (from 26 to 17) and jobs (from 120,000 to 80,000).

This policy balances what the Germans perceive as a strong need to maintain security of energy supply against the fact that the average cost of German coal is twice that of UK coal. Thus the Coal Concept provides for the continued subsidising by federal and regional governments of coal purchased by the German steel and electricity generation industries. Meanwhile, job losses will be offset by voluntary redundancies, job transfer and retraining, and regeneration packages for areas affected by the closures.

As McAvan states in her conclusion:

These energy policies and pit closure programmes have been developed within long-term frameworks which recognise that industrial reconversion requires at least 15 years on average if it is to be at all successful. (McAvan, 1993, p. 173)

She adds that the governments concerned have been rewarded by 'greater public confidence and better prospects of recovery' (p. 174).

British academics have also referred to German regional innovation strategies as a possible model for industrial and cultural regeneration in their own country. In arguing for a strategy 'that can begin to chart a more rewarding form of economic development in the future', Morgan and Price (1992) advocate that policy-makers in South Wales try to emulate the successful example of Baden-Wurttemburg. This particular industrial region is said to owe its success to a 'robust regional innovation strategy' characterised by: a firmly established research and technology transfer system which affords local firms access to advanced technologies and endows them with assistance and advice; a top quality vocational and educational system geared to local labour demand; a strong 'networking culture' connecting local buyers and suppliers; a collaborative relationship between the public and private sectors on matters of regional development; and a high degree of self-governance, ensuring that all policies (whether cultural, educational or involving training and technology) are tailored to the needs of the local region.

Another highly self-governing region, the famous Ruhrgebeit, has demonstrated a similar capacity for innovative industrial restructuring based on massive investment from government and commerce. Its enlightened educational, cultural and environmental strategies have given rise to a proliferation of higher education establishments, enviable land reclamation schemes and the assiduous promotion of tourism (*Financial Times* 12.6.91).

While Baden-Wurttemburg has been justifiably nominated as a possible model for the rejuvenation of former British coalfield areas, it must be recognised that part of its success is owed to the fact that the region embraces the European headquarters (and key industrial and service centres) of leading multinational companies, such as Daimler-Benz, IBM, Kodak and Bosch (Harvie, 1994). Clearly, such a highly conducive situation is not easily reproduced elsewhere.

Nevertheless, the underlying principles of this characteristically German approach have been tacitly endorsed in the UK, notably in the form of the so-called Neath Declaration, an ambitious agenda for the socio-economic rejuvenation of the South Wales valleys. This declaration envisages a reconstituted democratic structure based on an 'elected Welsh assembly and strengthened local councils', and sets out to guarantee 'high quality education and training to counter the geographic isolation, class and gender inequalities and low expectations of valley communities' (Baxter, 1992, p. 13). State intervention is seen as being required to promote the concept of 'industrial villages', based on small and medium-sized firms, to which a partnership of industry and national and local government would try and attract capital investment from Europe.

Morgan and Price state in their report that, 'In contrast to the UK approach, where innovation is left entirely to market-forces, the German approach recognises innovation for what it is, namely, a collective social endeavour' (1992, p. 41). What is presently happening in South Wales (and this could easily be said of other major UK industrial areas) is that numerous institutions are pursuing their own parochial ends. What is actually required is 'a partnership approach, a disposition to collaborate to achieve mutually beneficial ends. In this context knowledge circulates much more rapidly in the regional system and the best practices of one part are broadcast to all' (1992, p. 44).

If Morgan and Price are correct, non-political alliances of this nature, epitomising Beynon *et al.*'s progressive approach to community politics, may represent a viable alternative for Britain's beleaguered mining areas. Regional development initiatives, tied to the economic interests of specific localities, and sensitive to local political and cultural traditions, could alleviate the worst effects of Britain's pit closure programme. Yet their introduction may depend on the support of a national government which still seems preoccupied with its own political agenda, regardless of the effects it may have on the economy and culture of ex-coalfield areas.

References

Allen, V.L. (1981), *The Militancy of British Miners*, Shipley, The Moor Press.

Barnsley and Doncaster TEC. (1992), Action Plan – *Dealing with Closures in Barnsley and Doncaster Barnsley*, Barnsley and Doncaster Training and Enterprise Council, November 1992.

Baxter, S. (1992), 'Valleys in the shadow of coal', *New Statesman and Society*, 11 September, 12–13.

Beynon, H., Hudson, R. and Sadler, D. (1991), *A Tale of Two Industries: Contraction of Coal and Steel in the North East of England*, Milton Keynes, Open University Press.

Coalfields Community Campaign. (1993), *Memorandum on the State of the Coal Industry*, Yorkshire.

Crick, M. (1985), *Scargill and the Miners*, Harmondsworth, Penguin.

Critcher, C., Dicks, B. and Waddington, D. (1992), 'Portrait of Despair', *New Statesman and Society*, 23 October, 16–17.

Dennis, N., Henriques, F. and Slaughter, C. (1969), *Coal Is Our Life: An Analysis of a Yorkshire Mining Community*, London, Tavistock (second edn).

Department of Trade and Industry. (1993), *The Prospects for Coal: Conclusions of the Government's Coal Review*, Cm. 2235. London, HMSO.

Dicks, B., Waddington, D. and Critcher, C. (1993), 'The quiet disintegration of closure communities', *Town and Country Planning*, 62(7), 174–6.

Edwards, C.Y. (1991), *Restructuring the European Community Coal Industry: A Study of the Social Consequences for the UK Mining Areas*, Kingston Business School Occasional Paper Series. Kingston, Kingston University.

Employment Service. (1992), *Local Action for Employment*, London, HMSO.

Fine, B. and Millar, R. (eds) (1985), *Policing the Miners' Strike*, London, Lawrence and Wishart.

Fothergill, S. (1993), 'The coal industry after the White Paper', *Town and Country Planning*, 62(7), 169–70.

Fothergill, S. and Guy, N. (1993), *The End of Coal? The Impact of the 'Dash for Gas' in UK Electricity Generation*, Barnsley, Coalfield Communities Campaign.

Fothergill, S. and Witt, S. (1990), *The Privatisation of British Coal*, Barnsley, Coalfield Communities Campaign.

Francis, H. and Smith, D. (1981), *The Fed: a history of the South Wales miners in the twentieth century*, London, Lawrence and Wishart.

Glyn, A. (1989), 'The economic case against pit closures', in *The Case for Coal*. Barnsley, Coalfield Communities Campaign.

Guy, N. (1994), *Redundant miners survey: Vane Tempest*, Barnsley, Coalfield Communities Campaign.

Hall, T. (1981), *King Coal, Miners, Coal and Britain's Industrial Future*, Harmondsworth, Penguin.

Harvie, C. (1994), *The Rise of Regional Europe*, London, Routledge.

HM Government. (1993a), *British Energy Policy and the Market for Coal*, House of Commons Select Committee on Trade and Industry. London, HMSO.

HM Government. (1993b), *Employment Consequences of British Coal's Proposed Pit Closures*, House of Commons Select Committee on Employment. London, HMSO.

Hudson, R. and Sadler, D. (1987), 'National policies and local economic intitiatives', *Local Economy*, 12(2), 1–14.

Jones, D., Petley, J., Power, M. and Wood, L. (1985), *Media Hits the Pits*, London, Campaign for Press and Broadcasting Freedom.

Lawson, N. (1992), *The View from No. 11*, London, Bantam Press.

Little, M. and Reynolds, S. (1991), *Guidance, Training and Education: Provision for Redundant Miners from Mardy Colliery*, Ammanford, West Glamorgan, The Valleys Initiative for Adult Education.

McAvan, L. (1993), 'There is an alternative', *Town and Country Planning*, 62(7), 171–4.

Miller, S. (1985), 'Grimethorpe now', *London Review of Books*, 7(10), 6 June 1985, 16–17.

Milne, K. (1993), 'Keeping the coal fires burning', *New Statesman and Society*, 22 January 1993, 12–14.

Morgan, K. (1993), 'Revival in the valleys?', *Town and Country Planning*, 62(7), 177–9.

Morgan, K. and Price, A. (1992), *Rebuilding Our Communities: A New Agenda for the Valleys*, London, Friedrich Ebert Foundation.

Morris, J. and Wilkinson, B. (1993), *Poverty and Prosperity in Wales: An Analysis of Socio-Economic Divisions*, Cardiff, Cardiff Business School, University of Wales.

Parker, M. and Surrey, J. (1992), *Unequal Treatment: British Policies for Coal and Nuclear Power, 1979–92*, Brighton, University of Sussex Science Policy Research Unit.

Parkinson, N. (1992), *Right at the Centre*, London, Weidenfeld and Nicolson.

Parry, D. and Gladstone, B. (1993), 'Pit closures in the Yorkshire coalfield: local responses', *The Regional Review*, 3(1), 7–9.

Rees, G. and Thomas, M. (1989), 'From coal miners to entrepreneurs: a case study in the sociology of re-industrialisation', paper presented to the British Sociological Association Annual Conference, Plymouth.

Samuel, R., Bloomfield, B. and Boanas, G. (eds) (1986), *The Enemy Within: Pit Villages and the Miners' Strike of 1984–5*, London, Routledge and Kegan Paul.

Scraton, P. (1985), 'From Saltley Gates to Orgreave, a history of the policing of recent industrial disputes', in B. Fine and R. Millar (eds), *Policing the Miners' Strike*, London, Lawrence and Wishart.

Taylor, A. (1984), *The Politics of the Yorkshire Miners*, London, Croom Helm.

Trotman, C. and Lewis, T. (1990), *Training and education: The experiences and needs of redundant mineworkers at Cynheidre and Betws collieries in South Wales*, Ammanford, West Glamorgan, The Valleys Initiative for Adult Education.

Turner, R.L. (1992), 'British Coal enterprise – bringing the "Enterprise culture" to a deindustrialised local economy', *Local Economy*, 7(1), 4–8.

Waddington, D.P., Dicks, B. and Critcher, C. (1994), 'Community responses to pit closure in the post-strike era', *Community Development Journal*, 29(2), 141–50.

Walker, P. (1991), *Staying Power*, London, Bloomsbury.

Wass, V.J. (1989), 'Redundancy and re-employment, effects and prospects following colliery closure', *Coalfield Communities Campaign Working Papers*, 5, 3–22. Barnsley, Coalfield Communities Campaign.

Wilsher, P., MacIntyre, D. and Jones, M. (1985), *Strike: A Battle of Ideologies, Thatcher, Scargill and the Miners*, London, Coronet.

Winterton, J. and Winterton, R. (1989), *Coal, Crisis and Conflict: The 1984–85 Miners' Strike in Yorkshire*, Manchester, Manchester University Press.

Winterton, J. and Winterton, R. (1993), 'Coal', in A. Pendleton and J. Winterton (eds), *Public Enterprise in Transition: Industrial Relations in State and Privatized Corporations*, London, Routledge.

Witt, S. (1990), *When the Pit Closes*, Barnsley, Coalfields Communities Campaign.

2

COAL POLICY IN GERMANY: HOW DURABLE IS THE MODEL OF CONSENSUS?

DR KLAUS SCHUBERT AND MARTIN BRÄUTIGAM

Introduction

Coal policy has been a continuously important topic of domestic and economic policy in the Federal Republic of Germany since the Second World War. This applies in particular to the area of hard coal, which is the focal point of the remarks which follow. The treatment of the subject is characterised by an easing of the tension between the comparatively dispassionate economic policy assessment, on the one hand, and an emotionally charged discussion of the future of hard coal, particularly in the regions concerned, on the other. The starting point for the economic policy discussion is the question of whether and to what extent coal mined in Germany is necessary as the basis for a secure energy supply. Depending on how this initial question is answered, the second group of topics, namely the question of coal-mining subsidies, is evaluated differently. Even if the need for an independent, domestic energy supply is rejected, current subsidies will perhaps be seen as justified for a transitional phase (the duration of which would have to be more closely defined). Those who represent this position are supported by the development of the European Union and the single European market, which makes the implementation of national subsidies significantly more difficult. In this connection, however, the question which then has to be answered is whether and to what extent the loss of jobs and economic power in the regions concerned can be mitigated by central government and other measures. Despite substantial changes in the economic and employment structure of the coal-mining areas, coal is still a factor which is barely replaceable in the region, at least in the medium term.

However, international developments in this decade also support the opposite position. The relative stability of the 1980s has given way to large-scale uncertainty in which, for instance, the long-term political developments in the major coal suppliers to the German market – Poland and South Africa – can no longer be forecast. If the role of German coal as a joint guarantor of German and European energy security is acknowledged, subsidies must be accepted as the longer-term solution, in view of the specific disadvantages besetting the initial position of German hard coal, particularly geological factors. It would also be conceivable for hard coal's position to be strengthened, as a result of the increasing sensitivity towards nuclear energy. However, this initial impression would appear to be challenged by the large-scale agreement among representatives of both energy types that the future of coal can only be guaranteed by maintaining nuclear power stations.

Nevertheless, there is a general consensus that there will have to be substantial cutbacks in overall coal output during the next few years. However, on the one hand, the position of the political forces which currently dominate is not clear and, on the other, the experience of recent decades has taught us that political objectives relating to target output levels have been continually changed. As a result, negotiations on the fixing of output levels are also becoming increasingly difficult, because the number of workers employed in the mining industry is also dropping in relation to falling output.

Until recently, the reduction in the number of employees in the mining industry had been largely absorbed by early retirement among the older mineworkers and the movement of younger workers from closed to working pits or by retraining them for other professions. However, this process has now reached its 'natural' limits, if only because there are only a few employees of pensionable age left. Making mineworkers redundant also appears far more problematic than with earlier adjustment schemes, in view of the unemployment figures in the coal-mining areas which are already well above average.

For the population of mining regions, the debate on the future of coal naturally means far more than a consideration of economic policy arguments. This applies particularly to those miners and their families who are directly affected; however, large parts of the remaining population are also taking part in the discussion on the future of hard coal. In the coal-mining areas in particular, coal policy has therefore always had a decisive influence on voting behaviour. This is indicated by the so-called 'coal group' in the North-Rhine Westphalian Parliament, which advocates the retention of hard coal across all party boundaries.

The questions which have only been briefly outlined here will be tackled and discussed in greater detail in the third part of this article. The second part firstly offers an overview of the history of German hard coal mining since 1945, in which the development of the consensus model of German coal policy becomes clear. In the next section, the pros and cons of the long-term conservation of hard coal-mining are discussed. Finally, in the fourth section, schemes are presented which are aimed at promoting structural change in the coal regions. In the closing summary, the future prospects for coal and the regions characterised by it are set out.

The development of German hard coal since the end of the Second World War

From 1946: coal as a mainstay of the 'economic miracle'

If one considers the situation in Germany immediately after the end of the Second World War, the superficial view is one of an industry which had been completely destroyed. Even up to 1944, industrial production had been continually increased as part of the wartime economy, with most of it finally collapsing during the last year of the war. A development comparable with that in industrial production was evident in the output of German hard coal. The rapid increase in output after 1946 was due to the fact that most of the pits were only badly damaged on the surface, while underground they had remained largely intact.

Between 1946 and 1957 hard coal output rose from 62 to around 150 million tonnes. During this time, domestic hard coal covered up to 70 per cent of the primary energy needs of the occupied zones and later the Federal Republic of Germany. A significant and still pertinent feature of hard coal mining in Germany is its close dependence on political decisions. Unlike other branches of the economy which developed freely in the social market system following currency reform, German hard coal output remained under the strict control of the Western Allies and later the European Coal and Steel Community. Coal was subject to government price-fixing and supply obligations. Even following the suspension of formal price-fixing by central government, coal was subject to 'a special price discipline associated with political pressure' (Federal Government Coal Commission, 1990, p.46).

The substantial degree of government control explains the value placed on coal during Germany's economic revival. The background to this key position was the consideration that in view of anticipated bottlenecks in supplies of primary energy sources to Western European countries, the safest precaution was to use domestic resources as comprehensively as possible. Germany's hard coal mining industry (being the largest in Europe) played a central role in these considerations.

The coal crisis

As early as the mid-1950s economic changes were becoming apparent which were to result in a serious coal crisis from 1958. During the course of this, two-thirds of employment in mining were lost, with 400,000 out of 600,000 jobs having been cut by the early 1970s. During the same period, coal output dropped from 150 million tonnes a year to 100 million tonnes and coal's share of primary energy consumption fell from 70 per cent to 20 per cent. The reason for this dramatic decline lay with international developments.

Although the international assumption up to then had been one of an energy shortage, this assessment became obsolete almost overnight. The international perspective was now one of a surplus of energy sources. The reasons for this were, firstly, the increasing exploitation and use of large stocks of mineral oil in Africa and the Middle East and, secondly, the supply of natural gas now emerging on the market. During this phase of sharp drops in energy prices, the price-fixing policy which had previously been applied to hard coal was no longer sustainable.

In addition to this came a shift in the dollar exchange rate, which made German coal significantly more expensive on the world market than energy sources traded in American currency. Furthermore, the mining companies were unable to react to the emerging structural changes. The consequence of falling energy prices coupled with new rival products was a long series of pit closures. The massive job cuts which resulted from this were largely made without social provision, such as social plans or early retirement. However, they were accompanied by retraining schemes and the vocational and geographical mobility of many former mineworkers. Due to the fragmentation of the hard coal mining industry into many small mining companies, there was no planning behind the pit closures to take account of regional economic conditions, for instance; instead they took the form of a wild 'pit demise'.

Formation of Ruhrkohle AG (RAG)

RAG is a private-sector company. The shareholders are firstly steel industry companies and secondly power generators. Despite its private-sector structure, RAG is almost exclusively dependent on government targets, following in the long tradition already mentioned. So, for instance, the Federal government's energy programme from 1973 specified that output should be reduced from approximately 97 million tonnes at that time to 83 million tonnes in 1978.

The formation of RAG in 1968, which united 26 of the 32 German mining companies and thereby concentrated over 85 per cent of Germany's coal-mining capacity in one company, marked the end of this development. This concentration made it possible for any further reduction in capacity which might prove necessary to be more effectively controlled and for pits to be selected for closure based on their efficiency compared with other pits. Furthermore, RAG was in a better position to ease the social consequences of pit closures through the aforementioned policy provisions. In addition, there was the largely new possibility of transferring mineworkers to pits which were still working.

Consolidation during the oil price wars of the 1970s

As early as 1974, the underlying political conditions changed fundamentally. During the oil price shock, the importance of hard coal as a national energy reserve was fundamentally reassessed and central government targets altered accordingly. Instead of a reduction in mining capacity, the aim was now for hard coal output to be maintained at roughly 94 million tonnes a year. In 1977, two years before the second oil price war, the so-called 'coal priority policy' was formulated. The setting of output levels was abandoned, but domestic hard and brown coal was to take priority for the purposes of energy supply.

Despite the parallel economic crises affecting the economy as a whole, as far as coal was concerned, this consolidation phase in the 1970s and early 1980s was a period of increasing investment activity and modernisation of mining technology. The number of employees remained constant at 200 000, though the average age of miners decreased and the level of their skill qualifications increased. This development, which ran counter to what was otherwise an economic crisis, once again underlines the close connection between hard coal mining and political objectives.

Century Contract and reduction in capacity

The signing of the 'Century Contract' in 1980, which fixed the amount of coal used for the generation of electricity until 1995 through guaranteed sales to power generators, indicates the Federal government's intention to move from the spontaneous political targets of previous years to a long-term perspective. The coking coal subsidy compensated German steel companies for the difference between the cost of German and imported coking coal. In 1983, during the steel crisis, it also became necessary to reduce coal production by 10 million tonnes by 1988, in line with falling steel production, and then stabilise it at a new level. In 1986 falling oil prices and the weakening of the dollar increased the difference between the price of domestic energy sources and imported coal and oil drastically, to the detriment of German hard coal. In a Coal Round in 1987, a reduction in output by a further 13 to 15 million tonnes by 1995 was agreed.

Even the first output adjustment would have required substantial efforts on the part of the mining companies, since it involved the loss of 30,000 jobs. The additional drop in capacity made it necessary for central government and the mining *Lands* to provide increased support for the adjustment process.

Uncertain future – the 1990s

In the 1990s hard coal once again found itself under increased political pressure. On the one hand, the steel crisis had resulted in a significant drop in demand from one of its most important customers while, on the other, increasing pressure was being exerted by the European Union for a reduction and long-term abolition of all coal subsidies. Political debate at home was also questioning to an unprecedented extent, either directly or indirectly, the continued existence of a German hard coal mining industry. The focus of these discussions was not so much a target framework for coal output, but rather the level of subsidies which could or may be given.

This debate led to the call from the mining industry for the underlying political conditions for hard coal mining to be confirmed or redefined at a new Coal Round. Although this Coal Round did not take place, the mining industry felt compelled to cut its production more severely than had initially been agreed (and thereby also its number of employees), in view of its declining sales to the steel industry.

Another consideration in the present situation is that energy and electricity companies are no longer prepared to stand by the formerly guaranteed bulk purchases of German hard coal, due to the price pressure on the European market. Another substantial reduction in output and employment in German hard coal mining is therefore imminent.

Summary – dependence on political targets determines development

The preceding description explains the close relationship between political targets and the development of German hard coal mining. This close relationship has historical roots in the rigid government supervision of mining in the nineteenth century and in the variable political situations of the twentieth century. This tradition returned after the Second World War, due to the supporting role played by coal in the process of German and European recovery after 1945. The consequence of this has been that the economic development of hard coal mining runs largely independently of overall economic development and in some cases displays opposite trends. It is true that, in the longer term, there is a tendency for political measures to move towards a reduction in mining capacity. However, the fluctuations in this process are so considerable that the desired social compatibility of the adjustments is being made considerably more difficult. In the 1990s even the previously undisputed basis for coal subsidies – coal's role as the national energy reserve – is a subject for discussion.

The changing political demands can be attributed to volatile international influences, probably the most important being the price of energy on the international market. There is little suggestion that these international influences will remain steady. However, the basic tendency for mining output to continue to fall may persist.

Discussion – must German hard coal mining be permanently maintained?

As already indicated in the introduction, the central issue dividing those involved in the debate on hard coal production in Germany is whether the objective is for mining to be retained in the long term or not. In the following, the arguments of the advocates of such a long-term retention are discussed.

Historical, psychological and technical reasons

As already mentioned earlier, the question of maintaining the German hard coal industry is one which rouses deep emotions, particularly in the affected regions and among the mineworkers themselves. An ever-recurring argument is that, due to mining's long historical tradition, it should not be abandoned. Something of particular importance in this line of argument is the central role played by hard coal in the reconstruction of German industry after the Second World War, when up to 70 per cent of the primary energy needs of the occupied zones, the Federal Republic of Germany, were covered by the mines of the Saarland and Ruhr.

This phase in Germany's history explains the particular self-confidence of those living in the coal areas, especially the mineworkers, and their perceptions of themselves. This feeling of being a sort of industrial elite, a driving force behind the whole economy in times of need, has still not quite disappeared today. If this status, which is at the same time a myth, is questioned, either through political comments or the reality of economic developments, the result to date has regularly been violent, emotional reactions in the coal regions.

Against this background, the comments of the traditionalists and those who refer to mining's historical role can be assessed in two completely different ways. On the one hand, they are not unconditionally justifiable based on 'rational' economic assessment criteria. The outstanding importance of a branch of the economy in the past does not, in itself, justify lasting subsidies. A sector of the economy which is now structurally deficient cannot be maintained by subsidies amounting to billions of deutschmarks a year, because the financial resources used for this would be better invested in future-oriented projects than in those aimed at preserving the past.

However, where decisions such as those relating to the conservation of the mining industry are concerned, political reasons have been and still are relevant, aside from economic factors. This approach must take account of the fact that mineworkers, their relatives and the regional environment represent a quite significant voting potential. If one also includes the immediate supply industry and companies indirectly associated with the employees concerned, this produces a quite substantial voter potential, at least in the mining regions, which is particularly sensitive to political statements on the future of mining.

It is no surprise, therefore, that the political parties represented in the North-Rhine Westphalia Parliament almost unanimously represent a policy aimed at the long-term maintenance of coal-mining, albeit with a reduced importance. There is no doubt that any other position would significantly reduce the electoral fortunes of the parties concerned. This repeated political confirmation – from the majority of the political players, at least – further strengthens the marked self-confidence of the mineworkers. This means that the importance of strictly objective, 'rational' arguments is reduced, substantially in some cases.

Another argument for maintaining hard coal mining is given less attention. The closure of individual pits or the suspension of coal-mining as a whole would mean that the subsidies previously paid to mines could be saved. However, the notion that a closed pit generates no further costs is incorrect. In order to prevent the tunnels and shafts from collapsing, substantial sums – equivalent to normal operating costs in a working pit – have to be spent on making the mine safe, even after it has been closed. Although these costs are certainly substantially lower than current subsidies, they still have a certain significance.

Economic policy reasons – maintaining the basis for structural change

Although output capacity in the mining industry has been substantially reduced in recent years, coal is today still one of the most important economic factors in the mining regions. In 1993, around 75,000 workers alone were still being directly employed in mining by Germany's largest coal company, RAG (*Westdeutsche Allgemeine Zeitung* 18.6.94). In view of the fact that unemployment was standing at 13.3 per cent throughout the entire Ruhr (compared with 8.1 per cent throughout the territory of the former Federal Republic of Germany) in May 1994, the loss of a large proportion of these jobs would have had quite significant social consequences.

If one also considers the uneven distribution of these jobs within the region, this impression is strengthened still further. In a few towns in the Ruhr, RAG is the largest direct employer. In addition, it has been empirically established that for each direct job in hard coal mining roughly 1.3 additional jobs are indirectly dependent on it (Federal Government Coal Commission, 1990). Furthermore, in the RAG sector, for instance, almost 25,000 people are employed outside mining (*Westdeutsche Allgemeine Zeitung* 18.6.94).

If one follows the results of Lehner, Schubert *et al.* (1988), hard coal mining and its industrial environment must be seen as an essential factor in the development of new industries with future-oriented technology in the Ruhr. This study contradicts the frequently repeated view that it is precisely the crusty monostructures of the coal and steel industry which are impeding structural change in the Ruhr. According to these findings, only the disintegration of these structures will create the chances and opportunity for new industries to create profitable jobs with a wealth of prospects.

The frequently quoted principle that structural change cannot be achieved through structural collapse is underlined by the study produced by Lehner, Schubert *et al.* (1988). The study points firstly to the direct and indirect employment effects of mining, but also to the considerable innovative potential of the Ruhr's mining industry and its supply industries, not only in its own but also in other branches of industry, such as electrical engineering, mechanical engineering and the construction industry. From this point of view, the elimination of the leading and support industry of hard coal mining as an investor and consumer of various goods and services would have quite significant negative consequences for the future economic development of the Ruhr.

Although this perception of mining playing a positive role in the structural change of the coal regions is shared by many, this argument can only be used to support the maintenance of this industry for a limited time. So, to a certain extent, the enormous coal-mining subsidies would buy the time necessary for the structural change in the Ruhr or the cushioning of its social and political impact. Nevertheless, according to this view, hard coal mining still only has a period of grace, the length of which has to be negotiated.

The same notion of a gradual phasing-out is behind the ideas being developed in the European Union. According to these, subsidies for hard coal mining should be degressively structured; in other words, they should be gradually reduced, based on an established timescale. Nevertheless, because for largely geological reasons German hard coal is always produced at costs which are above the world market price level, a phasing-out of subsidies would undoubtedly mark the end of hard coal mining.

However, this gradual reduction in subsidies, which would at first glance appear reasonable, also neglects other important factors which have a crucial effect on the price of coal. This does not depend solely on direct costs and the increased cost-consciousness of the mining companies. The deficits of the RAG and other companies, i.e. the level of the necessary subsidies, is directly related to the world market price for coal or, more precisely, the difference between this and production costs in Germany. In other words, even if today's costs were further optimised, the exchange rate fluctuations and prices of other energy sources would remain a significant influence which could not be controlled by the mining companies. To this extent it is at least questionable whether direct manipulation of subsidy levels would automatically control the level of output or whether it would distort the whole process.

Coal as a national and European energy reserve

The most important argument in favour of keeping German hard coal mining is therefore based on its position as guarantor of a secure energy supply. At first glance, such logic appears outdated and reminiscent of the days before the establishment of free world trade and international cooperation. However, even at the end of the 1980s, the memory of the previous decade's oil price wars could have overshadowed any excessive optimism in relation to assessing the security of energy imports and imports of energy sources. The consequence of this was a policy of stabilising the shares of different energy sources in total consumption ('the energy mix'). In the longer term, though, the stabilisation of hard coal's share of the energy supply only partially benefited domestic hard coal. Imported coal can be regarded as the actual winner of the policy aimed at protecting individual energy sources from crises of supply.

Advocates of long-term subsidies for German hard coal mining argue the reliability of domestic coal as an energy supplier. On the other hand – based on this line of argument – all imported energy sources appear in varying degrees to be susceptible to risk. In the past it has been shown time and time again that predictions on political developments, particularly outside Europe, have always had to be corrected. Advocates of continued subsidies for German hard coal point to possible uncertainties in the political development of Eastern Europe, for example (Poland should be mentioned here as an important coal exporter), but also South Africa, which mean that reliable predictions cannot be easily made. The maintenance of German coal-mining as a basic energy supply therefore appears to be the most valid reason for continuing to subsidise this industry in future too.

However, it is doubtful whether this view will win enduring political favour in the long term. What is more likely is a development which has become clear since the Second World War, which ties coal policy to the short- and medium-term assessment of the political and economic situation at the time.

Measures for structural change

Irrespective of the question of whether and to what extent coal will continue to be mined in Germany in future, the hard coal regions in particular have been heavily hit not only by the massive loss of jobs in mining, but also by the structural crisis in other industries, such as steel. Nevertheless, mining is still one of the most important employers and sources of economic output.

A summary is presented below of the measures taken by various players in the field of politics and the economy, in order to promote economic structural change in the coal regions, particularly the Ruhr. The threat of job losses in the coal and steel industry and its supply industry, in particular, necessitates substantial economic and political intervention. The following description is divided into four aspects: removal of mining damage, possibilities for business promotion, public/private partnership and independent efforts on the part of RAG.

Possibilities for business promotion

As already explained, from today's point of view any realistic scenarios for the future of German hard coal mining involve job cuts in the coming years to some degree. These will be accompanied by further job losses in the mining supply industries. Economic and therefore also employment losses would have a knock-on effect in, for instance, the retail trade.

Against the background of what is already very high unemployment in the coal regions, economic policy measures aimed at promoting structural change are necessary. The common thread running through the German contributions to this volume, to which this section serves as a brief introduction, is a new understanding of the role of the state and its players in economic policy. This is based on so-called non-hierarchical relationships, i.e. on co-operation with those affected in the economy and society. In view of the large-scale independence of economic players in relation to government influences, only this sort of partnership between government and the economy would appear reasonable.

This is made particularly clear by the analysis of Wulf Noll. In this volume Noll is concerned with the various schemes which the Land of North-Rhine Westphalia has implemented in order to support structural change in the Ruhr. The first of these projects to differ from the customary framework used up to then was the 1979 'Ruhr Action

Programme'. Apart from the infrastructural improvements which had been implemented to date and were still being pursued, an attempt was made here to bring knowledge of potential innovations to the companies concerned. So-called technology centres were set up for this purpose, which were to act as advisers and mediators. In 1987 another change was introduced in an attempt to deal more effectively with the specific requirements of each locality. The new business promotion programmes, such as the 'Future Initiative for the Coal and Steel Region', shift the decision-making on individual projects from *Land* government level to the regional conferences, which not only include local authorities, but also chambers of industry and commerce, trade unions and other important pressure groups. Noll's conclusion is that the basis for a successful structural policy can only be created through new forms of dialogue and co-operation between central government, institutions and companies.

In his discussion of the Business Promotion Company for North-Rhine Westphalia, Rainer Buhr presents one of the central institutions supporting structural policy. A focal point of the activities of this Land company is the procurement of foreign contacts for North-Rhine Westphalian companies. The aim is for this to be achieved primarily by presenting the *Land* and its companies abroad, in other words, by creating positive framework conditions for the activities of individual companies. What is particularly important with this type of marketing is that the information reaches the central decision-makers, the potential customers and business partners.

Another area in which the Business Promotion Company is involved is the co-ordination of company closures. At first glance, this would appear to have nothing to do with 'business promotion'. However, providing for the workers concerned and trying to create replacement jobs is also an important element of structural policy. A relatively new instrument in this respect is assistance for companies in the area of product development on new markets – Buhr quotes the example of the recycling of used cars in this respect.

Heiko Beenken introduces further aspects of the structural policy of North-Rhine Westphalia in his paper. For him, the focal point is active employment market policy. For some time now, one of the essential goals of all structural policy measures in the coal-mining areas has been to protect mineworkers from compulsory redundancy. Achieving this goal has become increasingly difficult as the speed with which mining jobs are cut has accelerated, and, in particular, since some of the hitherto tried-and-tested instruments of employment market policy are proving less and less effective. This applies especially to the movement of employees to other pits which are still working, the number of which is decreasing. Also, due to the widespread use of early retirement in the past and in view of the very young workforce (the average age in mining is currently less than 34), this instrument too can only be applied to a very limited extent.

Centre stage is therefore given to the attempt to create other vocational prospects for mineworkers using training schemes. Important in decision-making for such training programmes are the respective regional conferences (also emphasised by Wulf Noll). The content of the training courses is closely co-ordinated with those companies in the region who are seeking employees. This ensures that the training schemes do not become detached from the market and actual demand. These schemes are not financed exclusively by the Federal Department of Employment, i.e. the employment offices concerned, but also through funds from the European Union and mining companies themselves.

Roland Schäfer introduces the small Ruhr town of Bergkamen in his chapter. It is particularly affected by pit closure, since in 1991 almost half of its total workforce was employed in mining. Three groups of schemes supported by local authority policy measures are aimed at making structural change socially compatible. Apart from the allocation of new industrial space and land and additional general infrastructural schemes, the promotion of advanced training schemes is of central significance.

The advanced training of employees is beneficial, firstly, to companies, which can thereby satisfy their need for qualified workers. However, it is also particularly important for the workers themselves, who will have a significantly greater chance of remaining employed. Advanced training schemes in the town of Bergkamen are sponsored by private institutions belonging to the Technical Control Board, which is primarily involved in the advanced training of technical and managerial employees from industry, trade, services and authorities. In addition, one of Ruhrkohle's subsidiary companies has the job of giving advanced training to mineworkers. On the other hand, institutions are sponsored by the local authority; these extend from the general range on offer at the secondary school to special establishments which concentrate particularly on promoting employment for women. In Schäfer's view, co-operation between government and public institutions and private business initiatives is necessary, in order to guarantee the increasing need for demand-based advanced vocational training.

Ruhrkohle AG

Matthias Hessling presents the prospects for German coal-mining and the measures taken by Ruhrkohle AG, as Germany's largest and most important mining company, to support the structural change in the Ruhr. Hessling regards coal-mining in Germany as necessary in the long term too: firstly, to safeguard the national and European energy supply and, secondly, as an important basis for structural policy in the coal areas.

According to Hessling, this second aspect should not be taken to mean that by maintaining mining jobs, the current employment structure in the Ruhr is also to be maintained. On the contrary, he points to the fact that already today over 40 per cent of the group's turnover is earned outside the actual mining sector. RAG is involved in, among other things, the areas of power-processing, chemicals and, in particular, environmental technology. In view of the increasing political pressure on mining companies to increase cuts in production and therefore employment, this area of RAG's activities is becoming increasingly important in the interests of structural change and the avoidance of structural collapse.

Public/private partnership

The need for co-operation between public and private players, who must free themselves from their previous hierarchical relationships, has already been repeatedly touched upon in earlier comments. With the Campaign for the Future of Coal-Mining Areas (ZAK), introduced by Wolfgang Steingräber, an exemplary project is presented. ZAK is an amalgamation of mining towns and communities in the Saarland and Ruhr and also the National Hard coal-mining Association, i.e. the employers' association and the Mining and Energy Trade Union (IGBE).

ZAK is regarded as the national arm of the European organisation EURACOM, a coal pressure group with representatives from seven European countries. The aim of ZAK is firstly to maintain coal output in Germany and secondly to support structural change in the coal-mining areas. Only the maintenance of a modern coal and steel industry in the coal-mining areas can facilitate a forward-looking structural change. According to Steingräber's comments, ZAK's functions involve, for example, organising and providing seminars and other events at local authority level, in which the political and economic players and decision-makers can enter into an ongoing dialogue. The constant exchange of information and objectives not only makes for mutual understanding, but promotes internal stability and security, particularly in times of extreme uncertainty, and thereby paves the way for successful action strategies based on consensus.

Another institution which has emerged from the co-operation between public and private players is introduced by Franzis Roxlau-Hennemann as the Eastern Ruhr Development Agency (EWA). The main function of this institution involves the restoration of former industrial land, particularly land containing disused mines. There are several reasons why this process cannot be controlled by market forces. In the past, for instance, the coal-mining companies themselves had no interest in the settlement of new companies, as these were regarded as rivals on their own employment market. Now the main problem lies in attempting to stimulate demand for land previously regarded as environmentally despoiled. This situation was only improved through the buying-up and financing of the rehabilitation of such areas by the *Land* government.

The EWA undertakes more extensive co-ordination functions in this respect. The EWA is a limited company whose shareholders include not only four local authorities from the eastern Ruhr, but also chambers of industry and commerce from the region and important business enterprises.

Removal of mining damage

Mining activities are fundamentally associated with ecological and other forms of damage. This damage occurs, on the one hand, while the mine is working and includes, for instance, a drop in the ground-water level, mining damage to buildings, etc. On the other hand, there is also an impact in connection with mining operations which have already closed down, for example, waste products in the soil. The latter type of damage is particularly important to structural change in former mining areas and those still working. What are in some cases very large interconnected areas, left unused after the pits and associated industrial plant, such as coking plants, have shut down, can generally be used as new industrial land.

Apart from legal difficulties over rights of ownership, the problem of industrial waste products crops up again and again, in the form of soil contamination, disused buildings and, in some cases, very deep concrete foundations. As a rule, these waste products must be removed before the land can be put to a new use. Apart from the practical problems involved in implementing such measures, the question of who can and will assume the rehabilitation costs must also always be answered.

The Ruhr has already been a source of valuable experience in this sphere of work, which is important to structural change. Andreas Wieder's chapter, therefore, deals with the problems encountered in the mining areas of the five new Federal *Lands*, due to the peculiar nature of mining in the former GDR, which for political reasons had to work very low-yield deposits. Now, following reunification, the mining companies are having to tackle particularly severe economic and waste product removal difficulties. Coupled with pit closures by various unprofitable operations, this led to more than a quite significant decline in employment: the great push towards closures at the same time produced a drastic increase in the waste product problem. Rather than the workers concerned being made redundant, at least some of them were now able to obtain continued employment in so-called rehabilitation enterprises. These undertakings are financed for the most part by funds from the Federal Department of Employment which it would otherwise have had to pay out as unemployment benefit.

Friedrich Wilhelm Wagner is concerned with the practical problems associated with the rehabilitation of former mining land. Since as early as 1966 the German Coal-Mining Areas Action Group has been involved in the rehabilitation of these areas of land. By 1988 it alone was able to put roughly 17,000 hectares of industrial and other land to new uses. Wagner points out that former mining land is generally viewed with some scepticism by investors, since the risk of waste products is regarded as very high. In order to confront this reluctance, the waste product situation must be resolved before the official release and sale of the land concerned. In particular, any remaining risks must not become a burden to the new investors. Consequently, from a legal point of view, the situation regarding former mining land must be assessed differently from that of other property used for industrial purposes. Unlike other branches of industry, the mining companies are obliged to rehabilitate their former property before it can be put to any other use.

Measures which go beyond this legal obligation on the part of the mining companies have been introduced by the *Land* of North-Rhine Westphalia. Wagner goes on to describe the measures needed for the far more serious ecological effects of brown coal (open-cast) mining. He refers to the strict statutory regulations governing the procedures before, during and after the working phase. In order to secure the specified objectives, the recultivation planning must start early and follow the mining work step by step. The restoration of the landscape relates not only to its external appearance, but goes as far as the specific soil structure which must be in line with regional characteristics.

Summary and outlook

It is certainly no exaggeration to describe the current situation of German hard coal mining as crucial for its future. However, due to the nature of consensus surrounding German coal policy, as described above, it is hardly conceivable that this fundamental question will receive an equally fundamental resolution. What is far more likely is a continued 'muddling through', but with the continued trend towards a reduction in mining capacity. To this extent, the circumstances described earlier in this chapter are likely to heavily influence the rate at which the mining capacity is reduced. Consequently it is difficult to estimate the extent to which hard coal mining will remain a mainstay of the economy in coal-mining areas in the long term or its potential contribution to structural change in the local economy.

The importance assigned to the longer-term conservation of hard coal for the economic development of the Ruhr becomes particularly clear if one looks beyond this industry to the other important branches and sectors of the Ruhr's economy. Here, there are many branches of industry suffering from similar structural crises to mining. Particular emphasis must be placed on examples from the steel and textile industries. The decline of these branches of industry, along with mining, has produced an extremely difficult situation for the Ruhr's economy overall. There is a great danger that the old structures may collapse too quickly and the creation of new structures may be too slow in taking off.

Against this background, the political processes underlying the necessary structural change appear even more essential. The common thread running through almost every German contribution to this compilation is a new understanding of the nature of co-operation between central government, the economy and pressure groups. Almost independently of the political colour of the governments concerned, this is characterised by a partnership of co-operation between public and private players. The fact that all sides are convinced of the need for joint action and the crossing of the gulf between central government and private companies, on the one hand indicates that the situation is regarded as extremely serious and a potential threat to the existence of the Ruhr as an economic area. On the other hand, this co-operation can also be seen as an indication of the determination of all those involved to offer the coal-mining areas an economic, social and cultural future, even if coal becomes less important.

References

Andersen, U. and Langmann, A. (1990), *Regionaler Strukturwandel und Regionalpolitik im europäischen Vergleich*, Bochum: (asp).

Coal and Steel Regions Commission of the Land North-Rhine Westphalia (1989), *Report*, Dusseldorf.

Coal and Steel Regions Commission of the Land North-Rhine Westphalia (1989), *Volume of Tables*, Dusseldorf.

Federal Government Coal Commission (1990), *Interim report* (of the Mikat Commission), Essen.

Future Campaign for the Coal Regions (ed.) (1991), *Resolution zur Kohlepolitik*, der Städte und Gemeinden, Hamm.

Klemmer, P. and Schubert, K. (eds) (1992), *Politische Maßnahmen zur Verbesserung von Standortqualitäten*, Berlin.

Lehner, F. and Schubert, K. *et al.* (1988), *Die Bedeutung der Ruhrkohle-AG für die wirtschaftliche Entwicklung des Ruhrgebietes*, Working Group for Applied Social Research and Practical Advice (asp), Bochum.

Lehner, F. and Schubert, K. *et al.* (1989), *Probleme und Perspektiven des Strukturwandels der Bergbau-Zulieferindustrie*, Working Group for Applied Social Research and Practical Advice (asp), Bochum.

National Association of German Hard Coal Mining (1991), *Steinkohle 1991*, Daten und Tendenzen, Essen.

Petzina, D. (1990), *Das Ruhrgebiet im Industriezeitalter*, Cologne.

Schubert, K. (1989), *Die Bedeutung des Steinkohlenbergbaus und der Bergbau-Zulieferindustrie für die Entwicklung des Ruhrgebietes*, in Glückauf 7/8, vol. 125, 1989.

Schubert, K. (1990), 'Standort Bundesrepublik Deutschland, Eine politisch-ökonomische Analyse', in Josef Schmid and Heinrich Tiemann (publishers), *Aufbrüche: Die Zukunftsdiskussion in Parteien und Verbänden*, Berlin/Marburg.

Schubert, K. (1991), *The Ruhr area's economic links with the coal industry*, Submission to the Committee on Energy Research and Technology of the European Parliament's Public Hearing on Coal Policy, Brussels, 18 January 1991.

Schubert, K. (1991), *Structural change and the impact of the mining supply industry*, Submission to the Committee on Energy Research and Technology of the European Parliament's Public Hearing on Coal Policy, Brussels, 18 January 1991.

Schubert, K. *et al.* (1991), *Dezentralisation und Kooperation: Elemente weicher Wirtschaftspolitik bei der Restrukturierung des Ruhrgebietes*, Bochum: Zentrum für interdisziplinäre Ruhrgebietsforschung der Ruhr-Universität Bochum (working paper 1/1991).

Westdeutsche Allgemeine Zeitung dated 26.4.91, Der Bergbau wartet auf neues Energieprogramm, Essen.

Westdeutsche Allgemeine Zeitung dated 7.8.92, Brüssel geht Europas Bergbau an den Kragen, Essen.

Westdeutsche Allgemeine Zeitung dated 4.1.94, Brüssel steht 1994 vor neuen Problemen, Essen.

Westdeutsche Allgemeine Zeitung dated 18.6.94, Die Ruhrkohle AG driftet in zwei Teile, Essen.

PART II

INDUSTRIAL REGENERATION THROUGH INNOVATION, CONVERSION AND DIVERSIFICATION

3

THE ROLE OF RUHRKOHLE AG IN GERMANY'S ENERGY AND COAL POLICY

DR MATTHIAS HESSLING

Introduction

In the paper which follows I would like to try to explain what the long-term changes in basic energy policy conditions have meant, and continue to mean, for German hard coal mining and the company Ruhrkohle AG. In particular, I would like to look at how the massive reduction in hard coal production and cuts in workforce have been handled to date, how they will be handled in future and what impact this is having on the changing economic structure of the Ruhr. One subject which will be given particular consideration will be the diversification policy which Ruhrkohle AG has been successfully pursuing for some time.

However, I would also like to deal briefly with the basic energy policy conditions themselves. These have meant that German hard coal mining, despite its high production costs compared with the current world market price for coal, is not having to prepare itself for a complete run-down. On the contrary, its long-term future has, in fact, been secured, albeit on a smaller scale than at present. This right to a long-term existence relates to the role given it by the Federal government in its energy policy as a contribution to ensuring a reliable energy supply; in other words, to guaranteeing a basic level of security in energy supply.

This long-term perspective means that Germany's starting conditions are entirely different from those in hard coal mining in other EC countries when it comes to the regeneration of coalfield areas. But this does not mean that the problems which have to be overcome are any smaller, as will become clear shortly.

Basic energy policy in the immediate post-war period

To provide a better understanding, I will begin by briefly illustrating the volatile historical development. For this, let me go back to the end of the Second World War. At that time the Ruhr's economy was in ruins and hard coal mining had declined to only a quarter of its 1938 level. It was not long, though, before the importance of Germany's hard coal to the regeneration of the region and of the company as a whole became evident. The Korean Crisis of 1950–53 and the resulting energy shortage highlighted its importance. Hard coal mining experienced a sudden boom.

Output reached its peak in 1957. At that time, 141 collieries in the Ruhr alone were producing 123 million tonnes of hard coal, while coke production had reached 40 million tonnes. The total number employed in the Ruhr's mining industry fell just short of 500,000.

However, even during the economic boom being experienced by the Federal Republic of Germany at this time, to which German hard coal had made a significant contribution, a dramatic change occurred in the economic situation of German hard coal mining. Germany's return to the world markets had been accompanied by a revaluation of the deutschmark against the US dollar, which meant that the country could 'afford' to import more energy. In the ensuing period, cheap mineral oil from the Arab states, sometimes at cut-throat prices, also pushed its way to the front of the heating market. Until then, this had been the traditional sales territory for hard coal. Furthermore, diminishing sea freight costs also made imported coal cheaper and the specific coal consumption of steel and power production dropped as a result of technical improvements.

As a result of these developments, hard coal output in the Ruhr had fallen from 123 to 90 million tonnes by 1967. Its workforce had shrunk from almost 500,000 to fewer than 230,000. Pit closures and the rapid reduction in workforce were often uncontrolled and uncoordinated. Then, in the second half of the 1960s, came the realisation that such a dramatic structural change in the energy market could not be accomplished by simply using the tools of the market economy. It became clear that an excessively rapid and unregulated adjustment process would inevitably lead to an economic crisis.

The creation and early activities of Ruhrkohle

What was needed was an orderly reduction in capacity and concentration in optimally sized businesses, taking social and regional economic interests into account. These ideas were embodied in the Law on the Adjustment and Regeneration of German Hard Coal Mining and the German Hard Coal Regions, which came into force in the middle of 1968. Central to this was the reorganisation of hard coal mining in the Ruhr with the formation of Ruhrkohle AG in 1968. Twenty-six individual companies with 52 collieries and 186,000 employees merged their mining assets to form Ruhrkohle AG.

In August 1970 Ruhrkohle AG presented a fundamental programme aimed at achieving concentration and adjusting production capacity to sales potential. It still applies today, although the targets have been modified several times since then. Adjustment as defined by this fundamental programme means:

- optimum allocation of deposits to the individual collieries and concentration on the most profitable collieries;
- no job losses as a result of closures or rationalisation measures;
- timing and local distribution of adjustment measures, to enable staff transfers and socially acceptable staff reductions;
- priority given to mergers over total closures;
- planning and operation of satellite mines rather than completely new ones with environmental conservation in mind; and
- consideration of regional balance, particularly with regard to jobs and training places.

This concept has enabled Ruhrkohle AG to fulfil its objective to date. The result is a hard coal mining industry which has diminished in size but is more efficient. Expressed in figures, since 1970 Ruhrkohle AG has reduced the number of collieries from 52 to 15 and the number of coking plants has dropped from 25 to four; while the number of briquetting plants has dropped from five to only one. The saleable output has been reduced to 46 million tonnes from its former level of 85 and coke production from 26 to 6 million tonnes. These adjustments in capacity have led to the workforce being reduced from 186,000 to 79,000.

At the same time, Ruhrkohle AG has remarkable successes to show in the technical field, in rationalisation and mechanisation. Our mines are now among the most modern in the world. This has meant that the average daily production per mine has risen from around 6,500 tonnes to 12,000 tonnes since 1970. At the same time, production in the mines has been concentrated on the best deposits, creating the conditions for full mechanisation and leading to a sharp rise in output per working unit. All in all, this has made it possible for the number of coal-faces to be reduced from 360 to 88, while the daily production per face has been increased from 890 to almost 2,000 tonnes. This led to a rise in productivity from 3,846 to 5,082 kilos per man-shift in 1992.

Energy policy in the 1970s and early 1980s

This technical development has still been hindered by the fact that the basic energy policy conditions have changed several times during Ruhrkohle AG's relatively short history. Up to the start of the 1970s, Ruhrkohle AG had focused on reducing its production and workforce. However, after the first oil crisis in 1973 and the economic loss which ensued, there was a reassessment of domestic energy sources and German hard coal in particular. The objective was to stabilise output. As a result of the second oil crisis of 1979, Germany's hard coal mines were even asked to increase production. The '90/90' formula was created, meaning that in 1990 Germany's hard coal mines were to produce 90 million tonnes.

This political goal was later abandoned under the effects of an apparent easing in energy supplies and prices. In the mid-1980s, in particular, the American dollar dropped sharply, losing almost half its value against the Deutschmark between 1985 and 1990. Furthermore, prices also fell on the world energy markets due to the divergence of interests within the OPEC cartel.

Finally, the temporary over-capacity on the world coal market, which is still with us today, meant that exporters had to struggle simply to maintain a market share by means of ever-decreasing prices. The result is that today companies are often forced to charge less for exports than their actual production costs, thereby producing a price level at which the investment urgently required to develop production and transport capacity, in order to enable them to respond to the rise in demand which can be expected in future, is neglected.

These energy price trends meant that, despite the fact that German mining companies' costs had remained constant in real terms, the cost disadvantage suffered by German hard coal producers compared with the price of energy imports, particularly imported coal, became increasingly severe and with it grew the need for subsidies.

In this environment of growing political and economic pressure on German hard coal, an agreement had to be reached at the 1987 Coal Round on accelerating the adjustment of output and workforce.

However, the enormous financial commitment involved in German reunification led to a new Coal Round as early as 1991, even before the measures agreed in 1987 had been completed. The result of this Coal Round was a further reduction in the German hard coal mining production capacity by 20 million tonnes within a few years, something which would mean the loss of 40,000 jobs in mining alone and 100,000 if we consider associated industries.

Nevertheless, the Federal government has always made it unmistakably clear that the long-term existence of a German hard coal mining industry is essential from the point of view of energy policy. In this way, the Federal government has given, and continues to give, German hard coal mining its actual *raison d'être*, by instructing it to make a reliable contribution to the country's energy supply. In other words, German hard coal is to contribute to the security of supply.

In this respect, the emphasis placed on security of supply in German energy policy must be viewed in the context of an increasing dependence on imported energy supplies and the risks inherent in this. Even today, Germany's dependence on imported energy is very high, standing at over 50 per cent. It is assumed, however, that this figure will already have risen to 63 per cent by the year 2000. The risks attached to such a dependence on imports provide more than enough material for a chapter of their own. However, they are not part of the subject matter being dealt with here. So let me just say the following in summary.

It is impossible to make an accurate prognosis of future trends on the world energy and coal markets. However, there are evident risks attached to the adequate supply of energy, including hard coal, on world markets at acceptable prices. German hard coal limits the extent to which the energy supply is exposed to the risks of the world energy markets. Although it does not provide an absolute guarantee, it makes a significant and reliable contribution to the German and, moreover, European energy supply.

If I have spoken about the role played by German hard coal in securing supply, it is because this constitutes the *raison d'être* of companies such as mine and is important to understanding long-term business functions, as well as the extent and aim of the restructuring and regeneration of coal-mining regions. Despite the emphasis placed on its importance in securing supply, under no circumstances should the social and economic significance of hard coal mining be neglected. Only in the very long term is there any possibility of changing regional economic structures, making the social consequences of restructuring tolerable and finding or creating alternative employment. Therefore, the social and regional economic significance of mining can only determine the speed at which an adjustment, in other words, a reduction in output and workforce, is possible; but the essential point is still the contribution to security of supply.

Ruhrkohle in the economy of the Ruhr

However, knowledge of the social and regional economic significance of hard coal mining is particularly important when it comes to establishing the speed at which an adjustment can take place (if it must for fiscal policy reasons). At this point, I would just like to give you a few figures on Ruhrkohle AG.

In 1992 Ruhrkohle AG paid out around 5 billion DM in wages and salaries and awarded contracts worth almost 7 billion DM, with by far the majority of the contracts (85 per cent) staying within North-Rhine Westphalia.

In addition to Ruhrkohle AG's workforce of just under 80,000, according to the calculations of an independent institute, there are also around 100,000 workers who are indirectly dependent on it, working in the mining supply industry, in companies specialising in certain mining jobs and in other areas of activity 'on the margins of coal'. Of these 100,000, by far the majority (80 per cent) are again employed in the Ruhr.

The Ruhr still has numerous towns and communities where mining provides a very large proportion of the total employment – in some cases over 40 per cent or even up to 50 per cent – which is why the closure of a mine would have particularly serious consequences there. Naturally, we cannot insist on a region's employment structures being entirely impervious to change in the long run. The Ruhr is in the middle of a structural change and Ruhrkohle AG is taking an active part in this process. A few figures illustrate what has taken place in the Ruhr over the last two decades. Total employment in the region stood at 2.2 million in 1990, just as high as in 1970. However, the structure of this employment has clearly changed and the number employed in manufacturing has dropped sharply – in mining alone, as well as in the iron and steel producing industry, it has fallen from 587,000 to 310,000. At the same time, the number employed in services has risen sharply from 881,000 to 1,181,000.

The diversification policy

Ruhrkohle AG is playing an active part in structural change and was once again specifically asked to do so in the 1991 Coal Round. The company's diversification is making a major contribution to this. When Ruhrkohle AG was founded, the founding companies merged only their mining assets to form the new company, while retaining other sectors. Consequently, Ruhrkohle AG's turnover came almost exclusively from the mining sector to begin with. In 1970 only 2 per cent of the group's turnover came from outside the mining sector itself.

However, it was soon recognised that hard coal mining has versatile know-how, a highly qualified workforce and numerous technical and logistical facilities which can be used profitably outside mining too. Accordingly, it was an obvious step to further develop the company 'around coal' and diversify it. In so doing, it was never the intention to abandon German hard coal mining; this was, and still is, to remain quite clearly the company's core area. However, new activities were developed around this core area. These are related to hard coal mining and were therefore able to profit (in the broadest sense) from their parent company which led the field of German hard coal mining. At the same time, the subsidiary and associated companies help their parent company, Ruhrkohle AG, both to remain economically viable and to support the structural change taking place within the region by creating new jobs or securing existing ones outside hard coal mining.

The first aspect which I mentioned should in no way be misinterpreted as a 'lust for power'. The fact is, however, that today Ruhrkohle AG's mining sector is dependent on a not inconsiderable transfer of profits from its subsidiaries and associated companies, without which it plainly could not exist. The reason for this is simply that the German subsidy system is structured in such a way that the total difference between the price of imported coal and the cost price of German hard coal is not repaid. Instead, German hard coal mining companies are left with a substantial cost element which is not covered by subsidies. This significant financial burden can only be managed by setting it off against profits from the subsidiary and associated companies.

Apart from this justification of diversification as a 'battle for the group's survival', a particularly important argument is the endeavour actively to participate in the structural change taking place in the Ruhr. This is achieved by creating new employment opportunities which are also, but not exclusively, used by former employees from the mining sector. However, it is also achieved by rounding off the range of activities in which the Ruhrkohle group is involved by buying in. Although this does not create new jobs, it enables existing ones to be secured, thanks to the injection of Ruhrkohle's capital, know-how and the effects of synergy.

With this pragmatic approach, Ruhrkohle AG has forged ahead with the diversification of its business activities since its foundation. Almost 40 per cent of the total group turnover is now earned outside mining itself. Hundreds of associated companies of different sizes, which I cannot deal with here, have contributed to this turnover. However, I would like to introduce the major subsidiaries to you in a little more detail. The selected companies also represent individual divisions, according to which Ruhrkohle AG's overall holdings are classified.

STEAG AG is the group's largest company in the 'Energy' division. STEAG plans, builds and runs hard coal power-stations. With additional activities in district heat supply, energy technology, disposal, nuclear energy (both service and trade), micro-technology and media and process engineering, the company works in innovative fields. In 1992 STEAG recorded sales of 3.6 billion Deutschmarks with a workforce of just under 4,700.

In the 'Chemistry' division, Rütgerswerke AG, Rütgers for short, is the major company within the group. Rütgers is one of the most important manufacturers of carbon-based raw materials in process chemistry and, at the same time, produces and manufactures plastics, as well as being involved in road building and the design of roofing systems and structural seals. This company's turnover was 4.3 billion DM in 1992 when it had a workforce of approximately 15,000.

Particularly active in the 'Environment' division is Ruhrkohle Umwelt GmbH. The company was founded in 1988 to combine the various environmental activities which Ruhrkohle AG had been involved in hitherto and further develop them based on a specific strategy. Its main areas of activity are environmental and refining technology, the rehabilitation of contaminated land, the waste industry and disposal from industrial plants, the recycling of residual substances and environmental logistics. With almost 1,100 employees, Ruhrkohle Umwelt recorded sales of over 500 million Deutschmarks in 1992.

Although there are several companies within the 'Mining Technology/Engineering' division, I will only briefly mention Meuwsen & Brockhausen GmbH (M & B) here. Along with its associated companies, M & B offers products and services in the fields of mechanical engineering and steel construction, electrical and control engineering and information and communications technology. In 1992 it was able to achieve sales of a little under 300 million DM with a workforce of over 900.

The largest subsidiary company in the 'Trade and Sales' division is Ruhrkohle Handel GmbH. This nationally and internationally active company deals in both solid fuels and mineral oil products and is now one of Germany's largest fuel-trading companies. Furthermore, the company is also active in the areas of service/coal processing (for mixed coal supplies) and building technology. Ruhrkohle Handel's sales were over 2.9 billion DM in 1992, at which time it had over 1,600 employees.

Of the remaining subsidiaries and associated companies, I would like to introduce you to two in particular which are of special significance to the structural change underway in the Ruhr.

The first of these companies is Montan-Grundstücksgesellschaft or MGG. MGG's general function is to prepare and sell industrial and commercial land for new uses, particularly industrial and commercial, as well as to construct and manage buildings. By mobilising land, MGG is therefore making an active contribution to structural change in the Ruhr. Further successes have been achieved with the reactivation of former mining sites. Working in close collaboration with the respective local authorities and the international building exhibition, IBA-Emscherpark GmbH, projects of supraregional importance have also been implemented. These have been based on cost-minimising, use-oriented rehabilitation concepts. Since the formation of Ruhrkohle AG in 1969 approximately 27 square kilometres, or 6,700 acres, have been sold. In 1992 MGG declared its willingness to make available for new uses 20 mining sites which were no longer needed in 12 towns in the Ruhr, covering a total area of some four square kilometres, approximately a thousand acres.

The second subsidiary which merits a particular mention is Ruhrkohle Berufsbildungsgesellschaft or RBG. RBG is Ruhrkohle AG's educational enterprise; it currently runs numerous retraining programmes for mining employees who then have to find employment outside the company, or even outside the group. Apart from this, in 1992 RBG also retrained approximately 6,500 individuals outside hard coal mining in 20 educational centres in East and West Germany. This was mainly for careers in industry and business, as well as the service sector. Sponsors included public authorities, chambers of commerce, economic associations and private industrial and business enterprises.

So now we have been introduced to some of Ruhrkohle AG's subsidiaries. I must also point out that Ruhrkohle AG plans to restructure its associated companies sector. The reorganisation concept envisages:

- retaining mining interests unchanged within Ruhrkohle AG;
- combining non-mining interests within a temporary holding company trading as 'RAG Beteiligungs-GmbH', which would be wholly owned by Ruhrkohle AG; and

- structuring existing interests within RAG Beteiligungs-GmbH by division.

For this purpose, specific interests with related products and market sectors are to be combined in three new divisional parent companies – RAG Technik AG, RAG Umschlags- und Speditionsgesellschaft mbH and RAG Immobilien AG. Ruhrkohle AG plans to use this restructuring to further develop its non-mining sector and secure it in the long term. The reorganisation means, in particular, that:

- clear, distinct group structures will be created;
- areas of responsibility will be more clearly defined;
- decision-making paths will be shortened;
- customer and market proximity will be improved; and
- better financing possibilities will be created on the capital market.

It must be clearly stated that in the future Ruhrkohle AG will still be heavily dependent on profit contributions from its associated companies. The supervisory board has already approved the restructuring. The public authorities are expected to grant the necessary authorisation shortly too.

On the subject of diversification, I would like to move straight on to two points for which we are often taken to task when our efforts to further develop the group and the successes we have achieved to date are discussed. The first relates to intercompany sales, in other words, the proportion of sales generated by our subsidiaries and associated companies through business dealings with the parent company, Ruhrkohle AG. It is quite normal, in this respect, for newly-founded companies in particular to be initially characterised by very high intercompany sales when a group is diversifying. In other words, these companies could not exist without orders from the parent company. As time goes by, however, the proportion of intercompany sales will drop significantly as the company finds its way in the open market. It is precisely this process which is evident among Ruhrkohle AG's associated companies too. The individual companies are at different stages of detachment from their parent company, while others were never actually dependent on orders from Ruhrkohle AG. All in all, however, the proportion of total turnover generated by the subsidiaries and associated companies which is accounted for by intercompany sales is relatively small, standing at just over 20 per cent.

The second point relates to the total number of workers employed outside the mining sector as a result of diversification. They account for

21 per cent of the Ruhrkohle group's total workforce. Unfortunately, this figure is significantly lower than the proportion of the group's total turnover accounted for by the same companies (just to remind you once again, this stood at around 39 per cent in 1992). However, this is entirely representative of a trend away from labour-intensive industries which is not just evident in the Ruhr. Newly-located industries rarely create enough new jobs in one or a small number of companies to absorb the large number of job cuts which will have to be made in hard coal mining. This makes it particularly difficult not only to generate a significant proportion of turnover outside mining, but also to create or secure a satisfactory number of jobs through diversification. Nevertheless, the Ruhrkohle group now employs a total of over 25,000 staff in areas outside mining and this figure is displaying a marked upward trend. Although the Ruhrkohle group itself does not know the exact percentage, it can at least be assumed that a substantial proportion of these workers employed outside mining are still working within the Ruhr region.

I think it is generally accepted that it is both rational and necessary for companies operating in an environment of structural change to diversify. Even the structural change itself must be seen as right and important for the region. However, it is clear that structural change can only take place gradually if it is to be successful. Abrupt developments must be avoided in any event. It is therefore essential for the social and regional economic importance of hard coal mining and the consequences of closures to be taken into account when the speed of adjustment measures in hard coal mining is being discussed. Such arguments have now found a firm place in deliberations on reductions in production and the workforce, which have resulted in longer-term perspectives for German hard coal mining being reformulated at the numerous Coal Rounds which have taken place in the meantime.

The future context: 'Coal Concept 2000'

The last Coal Round took place in November 1991 and saw both producers and consumers, as well as the trade union and Federal and *Land* governments, reach a consensus on the future use of subsidised German hard coal. The results of the 1991 Coal Round have led to the formulation of the so-called 'Coal Concept 2005', which I would like to outline briefly here.

The most important outcome is that German hard coal will have to make a long-term contribution to ensuring a reliable energy supply in the united Germany and within the single European market too. In other words, the need to maintain a German hard coal mining industry in the long term, as part of Germany's energy policy, has once again been established.

It was also agreed, however, that there should be a gradual reduction in subsidised sales of German hard coal and therefore subsidised production levels. What this means in general terms is that the amounts of subsidised hard coal sold to the two strategically important sectors of power generation and the steel industry will be reduced to 50 million tonnes by the year 2005, although the mining industry has promised that it will have reached this quantity by the year 2000. This signifies a relatively short-term reduction in German hard coal production by around 20 million tonnes. What this means in concrete terms is a gradual reduction in supplies to German power-stations to 35 million tonnes hard coal equivalent by 1997, but with constant supplies thereafter until the year 2005, and a gradual drop in supplies to the German steel industry to 15 million tonnes by the year 2000, also followed by constant sales.

The steel mills contract, which represents a means of guaranteeing sales and subsidies for German coking coal, is to be renewed until the year 2005. The Federal government has also promised to develop a follow-on provision to secure hard coal power generation contractually and financially after 1995 and to obtain approval of the Coal Concept 2005 from the EC Commission. The mining companies, for their part, have been asked to take an active role in the restructuring of the affected regions within their respective frameworks.

The Coal Concept 2005 means that for reasons of financial policy German hard coal will make a reduced contribution to Germany's energy supply, while the speed of the reduction can still be made regionally and socially tolerable. But the adjustment process agreed there must certainly be described as extremely painful. The number of pits is to be cut by nine to only 17 and a total of 40,000 jobs will have to be lost in mining alone – this figure may rise to 100,000 if associated industries are included.

In the meantime, although the basic statements contained in the Coal Concept 2005 still apply, it has unfortunately proved necessary to bring forward the adjustment measures agreed at the time. The reason for this lies particularly in the crisis within the German steel industry which has led to an exceptional drop in sales since the second half of 1992. Therefore, it had to be agreed in March 1993 to bring forward once again the mergers of collieries still outstanding upon implementation of the Coal Concept 2005. As a result, Ruhrkohle AG's technical adjustment measures will already have been completed by the start of 1994, i.e. three and a half years earlier than originally planned. However, since it must be assumed that coking coal sales will be lower than planned in the coming years, there will still be a production surplus of 3–3.5 million tonnes, despite the bringing forward of mergers at Ruhrkohle AG, and three options for an additional colliery closure are currently under discussion.

Furthermore, the bringing forward of the mergers of collieries, coupled with the existing production surplus, means that a further cut in the workforce will be required in the short term. This had already been cut by over 28,000 between the start of the adjustment programme initiated after the 1987 Coal Round and the end of 1992. In addition to this, the staff cuts to be made by 1998 according to the Coal Concept 2005 will now have to come significantly earlier.

Apart from the measures employed to date, retraining schemes are being run and new employment opportunities created outside mining. As a result, Ruhrkohle AG will already have reached its target figure from the 1991 Coal Round by the mid-1990s. The company is still anxious in this respect to avoid operational redundancies. However, these can no longer be entirely ruled out, whereas earlier job security could still be guaranteed.

Future policy measures

It goes without saying that we will do everything possible to make staff reductions socially tolerable in future too. The measures which have already been used to achieve this in the past, and will have to be used even more in the future, are:

- blocking permanent appointments as far as possible;
- reducing trainee appointments;
- early retirement of older employees;
- intercompany transfers within the mining sector;
- staff transfers within the Ruhrkohle group as a whole;
- retraining schemes for employees from the mining sector;
- shorter working hours in line with the general trend within the West German economy,
- large-scale cancellation of overtime; and
- non-producing days to reduce production without affecting employment levels.

The potential for early retirement is becoming increasingly small. This is because the extensive use of this measure in the past has meant that there are very few older employees who could actually be considered for early retirement now. The average age of Ruhrkohle's workforce in 1992 was only 35. On the other hand, there is definitely potential in theory for staff who are to be made redundant to be employed in other activities outside mining. In 1992 only 4 per cent of the 'production personnel' in the mining sector had no training. Apart from staff with mining training, there was, in particular, also a significant percentage of employees, 56 per cent in fact, who had learnt skilled occupations in mechanical engineering or electrical engineering or were qualified in other areas which are in demand outside mining.

In addition to this, however, in future employees will have to make increased use of the retraining schemes mentioned earlier. It must be stated quite clearly in this respect that the prospects of success of retraining schemes, and also the efforts of employees without any mining training to find work outside mining, are very heavily dependent on the economic development of the region and of the country as a whole. The fact that the whole of Germany is currently in recession makes the conditions for successfully reducing the workforce in a way which is socially tolerable particularly difficult.

However, with the help of the public authorities and many other institutions, some of which are represented by the other German contributors to this volume, we will still be able to say at least that every effort is being made and, as in the past, we will undoubtedly succeed in tackling the remaining job cuts in a way that is socially and regionally tolerable while promoting structural change in the region.

4

REGENERATING THE DEARNE VALLEY IN SOUTH YORKSHIRE

HEDLEY SALT

I will divide this chapter into four parts. First, a brief description of the Dearne Valley and the problems facing the area. Second, an explanation of our response to these problems including the formation of the Dearne Valley Partnership, and the role of City Challenge. Third, an overview of the progress we have made over the last year in regenerating the Dearne Valley. And, finally, I will highlight the key issues, as I see them, affecting the Dearne for the next few years.

The Dearne Valley

The Dearne Valley Partnership area constitutes some 20 square miles, whose boundary overlaps with Barnsley, Doncaster and Rotherham Metropolitan Boroughs. It is, or was, a coalfield area. The population is 80,000 and these people live in a series of separate townships. The towns grew where the mines were sunk, and for no other reason. There is poor road infrastructure between the towns, the coal historically being moved by rail. There is a good surrounding motorway network, but poor access to it, and poor links across the area.

The Dearne Valley was completely dominated by coal, 'King Coal', and the nine pits of the Dearne that were open in the early 1980s employed 11,000 men. There is now one mine, with 500 men employed, and this will close in about two years time – through exhaustion.

The problems of the coal-mining industry and its destruction are well documented, but can stand reiterating briefly. Mining itself does not readily lead to spin-off industries which, in part, explains the narrow industrial base. More than this, the then National Coal Board had a clear policy, aided and abetted by successive governments, of actually discouraging inward investors into mining areas. They did not want competition for the available male workforce. It regarded the labour supply as its own.

The environment has suffered badly in all mining areas, but the Dearne is probably the most polluted and despoiled coalfield area within the UK: slag heaps, subsidence and poisoned land, criss-crossed with now defunct railway lines. It was an environmental moonscape. We have started to reclaim land, so there are some signs of improvement.

In education, what tended to happen was that further education was provided by this great, monolithic, paternalistic entity, the Coal Board. The mass of the male population went down the pits to work the coal. Those who were a bit brighter got apprenticeships in the industry, and the brighter still, very occasionally, went on to higher education usually via the Coal Board, to become engineers and the like. Thus the culture of the area did not place great emphasis on education, particularly higher education. Even in the miners' spare time, the industry made provision. It provided the sports field, it organised the bands, it sponsored the choir. Even houses were frequently rented from the mine, or the Coal Board.

Even if British Coal now wanted to make good a lot of the devastating legacy of coal-mining, it couldn't. It is being stripped of its resources ready for privatisation, so a key potential pillar of support has now effectively been kicked away. British Coal do attempt to assist through British Coal Enterprise, but given its budget, and given the scale of the problem nationally, this assistance, though welcome and useful, is totally inadequate.

However the Dearne Valley has one great source of strength: an overwhelming sense of community. Its people look after each other; team work and support are as instinctive as breathing. It is worth pointing out that in times of full employment, serious crime was virtually non-existent in coalfield communities. This resilience in the face of adversity is the foundation stone on which the regeneration of the Dearne is being built.

Our response

Let me start by explaining why the Dearne Valley Partnership was set up. It was set up because, in view of the massive scale of the area's problems and the area's administrative complexity, we wanted a new way of bringing resources and expertise to the area. At the start it was a local authority initiative, a local response by the three local authorities (Barnsley, Doncaster and Rotherham) to the loss of identity and self-respect suffered by the community in the Dearne Valley.

As the initiative developed, there was the need to involve private and community sector interest in an active way. But we were also well aware of the Government's agenda. What we did not want was a standard solution imposed from outside. We did not want an Urban Development Corporation to ride roughshod over people's wishes. A UDC is a government-imposed, non-accountable body with very wide powers, including planning and other matters, that cut right across local authorities' legitimate areas of responsibility. By contrast, The Dearne Valley Partnership is a locally accountable way of harnessing the benefits of working together.

A bid was made by the three local authorities in their Urban Programme submissions to fund a study on the Dearne, and to suggest a way forward to regenerate it. Submissions were invited from various agencies for this work. The Coopers and Lybrand Deloitte/Sheffield Polytechnic submission was not only impressive in its clear grasp of the problems of the Dearne, it also had qualities dear to every Yorkshireman's heart: it was highly competitive and relatively low cost. The key recommendation of Coopers and Lybrand Deloitte/Sheffield Polytechnic was to set up the Dearne Valley Partnership, consisting of the three local authorities, the private sector and the community sector. All of this predated City Challenge, the Government's invitation to bid for funding for partnership programmes, so we already had a partnership in place and were able to secure City Challenge funding.

The role of the Partnership Board is to set the strategy. It has to address itself to the task of the strategic deployment of local and central government resources within the Dearne; to set the broad guidelines within which the detailed work can take place; and to set the climate for inward investment.

Our strategy

This is our mission statement:

The Partnership aims to regenerate The Dearne Valley for the benefit of all its people by bringing together businesses, local government and central government for the common purpose of creating sustainable employment and improving the social and physical environment.

The brief we have set ourselves is to achieve a 'supply side' revolution in the Dearne Valley. What was envisaged as a ten-year struggle will be condensed under City Challenge to a five-year action programme supported by real resources.

By 'supply side', we mean:

- reskilling the workforce: education and training;
- providing a plentiful supply of land and buildings for new enterprises;
- providing new infrastructure to make the area accessible; and
- tackling the environmental problems and despoilation which detract from its appeal to new industrialists.

Our objective is to create 4,000 jobs over the five years of City Challenge. To do this we must attract nearly £300 million of private sector investment on the back of the £37.5 million of City Challenge money. Seven hundred and sixty hectares of land must be reclaimed; 290,000 square metres of industrial and commercial floorspace built; 1,700 dwellings built or improved; the Dearne towns link road must be completed as well as the spine road into the heart of the Dearne Wath Manvers. The centrepiece of our strategy will be a higher education institution in the Dearne, and I particularly refer to the importance of research and development facilities that such an institution would have in attracting the right sort of new industry.

Progress in the first year

Obviously the economic climate has been against us. Precious few green shoots have been evident in the Dearne (or elsewhere in the country). Further pit closure announcements in the surrounding area have exacerbated the problems. The Government's botched attempts to overcome the problems of contaminated land have only succeeded in making the problem worse. Despite this, we are on target to achieve our objectives. The Board is a genuine partnership where differences are aired and decisions made. The Community Forum has developed a life of its own and is raising the capacity of the community to determine its own destiny and to start filling the vacuum left by the loss of the coal industry.

Some issues obviously have been easier to deal with than others. Marketing and land release have proved particularly difficult, though we are now about to bring forward an integrated plan on this issue. On the other hand, great strides have been made in plans for a new university college at Wath/Manvers, right in the heart of the Dearne Valley. This visionary project will be a part of Sheffield University and will bring the opportunity of higher education to the doorsteps of thousands of young people for whom education has been seen as 'not for them'. It will accommodate 500 students when it opens in 1995, building to 3,000 students by 2005. Already British Coal have released the land and funding packages are being put together by the University and their partners, Henry Boots, with support from the Dearne Valley Partnership.This is an exciting project which will help transform the prospects of the area for inward investment and help ensure that local people are able to share in the area's new prosperity.

Many other projects are also proceeding throughout the Dearne including:

- new small factories at Thurnscoe;
- a new ITEC at Mexborough;
- further development of the Elsecar Heritage Centre;
- many community projects;
- the completion of phase 1 of the Dearne towns link road;
- the private sector development of a business park at Wombwell;
- new housing association dwellings at Thurnscoe; and
- the development of a business school at High Melton College.

Nevertheless, the main challenges still lie ahead. In particular there are five key issues for the future, listed in the following section.

Key issues for the future

1. Funding for the university college. As you may have seen in the press, the Government have decided against increasing the number of students in higher education next year. It is critically important that the Higher Education Funding Council is able to fund the student places planned for the University College of the Dearne Valley in 1995/6.
2. Enterprise zone. The Government have announced that the Dearne Valley will be able to apply for an enterprise zone provided there are sufficient job losses in the area arising from pit closures. This will have a catalytic effect on the area's inward investment prospects.
3. Urban Regeneration Agency. The URA is expected to become operational in the autumn of this year. It will take over the administration

of the derelict land programme, city grants and the English Estates'
Factory Building Programme, all of which are fundamental to the
success of our strategy. We are seeking an early meeting with the
chairman of the URA, Lord Walker and his Chief Executive when
appointed.
4.   Contaminated land. The Government must resolve the damaging
     effect which its ill-conceived land register proposals have had on
     investor confidence in reclaimed land. This can only be done
     through the provision of some form of indemnity guaranteeing the
     suitability of reclaimed land for development.
5.   Economic growth. Above all our strategy can only succeed when the
     economy comes firmly out of this recession and confidence returns
     to business and the property market.

Conclusion

Subsequent to the resolution of the above issues, I am confident that the
Dearne Valley Partnership will demonstrate to potential investors that
the Dearne Valley is an area with a great future and we will lever in substantial amounts of investments that would otherwise pass us by. It will
demonstrate beyond any shadow of doubt that, given the money and
given the chance, local government in partnership with the community
and the private sector can deliver. Whatever happens we will give the
people of the Dearne back their self-respect. The Dearne will have a
brighter future.

5

MANUFACTURING SUCCESS: INDUSTRIALISATION POLICIES IN DERWENTSIDE IN THE 1980s

PROFESSOR RAY HUDSON AND DAVID SADLER

Introduction

The closure of the British Steel Corporation's Consett works in 1980 was an event of profound cultural and economic significance within the town and the surrounding Derwentside district (see Kearney, 1990, pp.67–117). In many ways, it symbolised the deindustrialisation of that district and brought Derwentside to the forefront of public debate (see Robinson and Sadler, 1984). Industrial decline had in fact been progressing apace for the previous two decades as colliery after colliery closed as a result of the policies of another nationalised industry (see Hudson, 1989, pp.127–92). These colliery closures went largely uncontested, in part because they were spread over time and space and in part because of promises of alternative jobs. But these earlier efforts to counter the collapse of coal-mining via the construction of an alternative branch plant economy had had limited effectiveness. Thus the not entirely unexpected closure of the Consett works, which despite previous employment reductions was still far and away the biggest single manufacturing plant in the District, raised the spectre of generalised economic and social crises in and around Consett.

At the same time, following an unsuccessful campaign contesting the steel works closure (Hudson and Sadler, 1983), Consett became the focus of a vigorous campaign to create a newly diversified local economy, to replace the jobs lost with new ones born of a flourishing enterprise culture in local small firms. Different institutional arrangements involving co-operation between local and central government and the private sector, represented most clearly in the creation of the Derwentside Industrial Development Agency (DIDA), and intended to facilitate the emergence of this new economy, were launched in a blaze of publicity. Reindustrialising Consett came to be seen as the touchstone against which a variety of dogmas and policies about local economic development could be judged. As a consequence, those promoting the image of a successfully transformed town and district were acutely sensitive to even gentle and constructive criticism of the actual pattern of economic and social changes that emerged there in the 1980s (for example, see Hudson and Sadler, 1989, pp.109–16).

Much of the evaluation of the success of such reindustrialisation policies hinges on assessments of the numbers of new manufacturing jobs created in Consett and Derwentside. Clearly, many of those involved with implementing those policies and in selling Derwentside as a success story so as to attract further investment and jobs had (and have) a vested interest in projecting the district in as favourable a light as possible. In part this involves maximising claims as to success in attracting new manufacturing jobs. At the same time there are a number of deficiencies and ambiguities in what little 'official' information there is that is publicly available.

Thus a major problem confronting any analysis of social and economic change in Derwentside in the 1980s is that there is considerable scope for confusion and uncertainty as to the extent of new manufacturing investment, company formation and employment growth in the district. There is no doubt that some net growth has taken place – but there is debate and uncertainty both as to its magnitude, its character and its proximate causes. What we first seek to do here, therefore, is to bring together the varying claims and sources of evidence that are relevant to these questions and to a degree try to reconcile them. We then extend the discussion beyond a debate about numbers of jobs to some of the more qualitative aspects of the new manufacturing economy. Clearly, it is of the greatest concern in relation to defining an appropriate policy mix that the district council could pursue in the 1990s – that the pattern of manufacturing investment and labour market change in the 1980s be accurately assessed.

Estimating the number of new manufacturing jobs in Derwentside

The most accurate estimates of manufacturing employment change in Derwentside district in the 1980s are those provided by the Censuses of

Employment, although these refer only to employees in employment and not the self-employed. What they reveal is an increase of manufacturing employment from 4,830 in 1981, to 5,500 in 1987 and 6,100 in 1989. This gives a net increase between 1981 and 1989 of 1,270 jobs, of which 850 are male and 1,070 full time. Subsequent estimates suggest that there were about 200 self-employed workers in manufacturing in 1989. This represents the most authoritative statement that can be made about manufacturing employment change.

This clearly pales into insignificance when set against the net decline of over 8,000 in manufacturing employment in Derwentside between 1978 and 1981. Although only a small fraction of the massive manufacturing job loss of those years has been replaced, in a decade of continuing national decline in manufacturing employment, this could still be seen as a very reasonable commendation of local economic policy in Derwentside. There are, however, considerable discrepancies between the *net* numbers of new manufacturing jobs revealed by the Annual Censuses of Employment data and the numbers of new jobs claimed by the district council and the DIDA. Such discrepancies raise very serious questions as to the effects and effectiveness, in its own terms, of the reindustrialisation programme. Claims as to jobs created may have to be understood not so much as a description of what has happened but as part of a process of place promotion and perhaps, in some cases, institutional self-justification.

The intention in what follows is to try to go a little beyond the debate as to the numbers of jobs created and examine the types of jobs, the sorts of companies that provide them and the sort of new economy that these reflect. While accurate estimation of fresh manufacturing employment growth is clearly of great importance, there are also other important aspects of change in manufacturing that ought to be considered. At this stage, data limitations preclude a full examination of these but, nonetheless, some comments are possible.

New business start-ups

It is claimed that, between 1980 and 1988, £50 million of public expenditure in Derwentside 'triggered over £70 million of private sector investment, involving more than 200 businesses' (Carney, 1988, paragraph 1.4). This suggests a picture of a growing, healthy economy based around vibrant private sector investment. However, it also paints a rather one-dimensional image. Derwentside District Council's (DDC) *Progress Review* (1988a) gives a summary of the effects of the DIDP in terms of births and deaths of new firms and relocations into and out of the district over the period 1979–1988 and of some of the changes associated with investment and disinvestment.

These data suggest that there are considerable risks involved in many assisted projects, with almost 40 per cent of those companies receiving assistance having failed or ceased trading in Derwentside (possibly moving prior to failure). There are, however, marked variations in survival rates between different types of projects: only 13 per cent of inward investments and 15 per cent of expansions failed to survive, compared to 44 per cent of start-ups. This reveals, not surprisingly, that the risks are particularly marked in the case of new firms – hence the limited growth of self-employment in manufacturing – and these differential survival rates also raise questions as to the sort of new manufacturing economy that is emerging in Derwentside.

It is also possible to disaggregate the data on firms assisted and still successfully operating by 31 March 1988 by sectoral type, though only to a limited degree. These data show that the DIDP has not been solely concerned with the encouragement of new manufacturing firms and jobs, although this has been the most (self)-publicised aspect of the *Programme*, especially by the DIDA. It raises questions about the relationship between the manufacturing and service sectors and also about the extent to which service sector 'start-ups' are represented in the firms assisted through the DIDP.

Four key factors attracting business to Derwentside

There is no doubt that, in the 1980s, Derwentside witnessed the vigorously publicised promotion of a 're-industrialisation strategy', centred around the existence of 'probably the best project package in the UK'. (*Derwentside Fact File*, n.d.). In general terms, it is possible to say that firms have been attracted to Derwentside as a result of some combination of: investment in infrastructure and the general conditions of production by the EC, central and local Government; provision of grants and loans directly to companies by these same organisations; labour market conditions, including direct wage subsidies; and environmental considerations. The DIDA laid particular stress upon its involvement in the preparation of Business Plans. The selling of Derwentside to inward investment stressed that the 'Agency has developed its own computer model for locational comparisons . . . . In-depth comparative analysis can be provided for a chosen project with reference to a specified set of European locations' (*Derwentside Fact File*, n.d.). Clearly, the emphasis was on demonstrating that Derwentside offered the greatest potential profits to mobile private investment in projecting the district in the international place market.

Infrastructure, industrial estates and advance factories

In the wake of the steelworks closure, there was considerable provision of new industrial estates and advance factories planned for Derwentside. Although the provision of new capacity did not proceed at the intended rate, nonetheless, by the end of 1987, 1,156,000 sq. ft. of new factory floorspace had been built in Derwentside, of which 70 per cent – 820,000 sq. ft. – had been built by English Estates (mainly in one location Consett No.1 Estate). This stock of 'modern' factory space was seen as a crucial pre-condition for the attraction of new firms, especially inward investors. But by the end of 1987 the vacant stock of such space was reduced to 182,000 sq. ft., of which 84,000 sq. ft. was allocated to pending tenants.

The shortage of modern industrial units reflects both supply and demand factors. In part, it reflected slippage on the development of the Greencroft Industrial Park, seen as the most important new industrial estate in Derwentside after the Consett No.1 (see also Hudson, 1989, pp.320–26). It also reflected the fact that ongoing developments for a further 106,000 sq. ft. of new factories at Hownsgill and Tanfield Lea South were already 79 per cent allocated. This led the Chief Executive of the district council to write to the Secretary of State for Trade and Industry (3 December, 1987) to argue for extra Government funds for English Estates to provide further factory space in Derwentside. He argued (unsuccessfully) that 'the lack of availability of appropriate new advance factory space is an overriding threat to the continuing ability to attract and sustain growing businesses'.

Grants and loans

Financial incentives have clearly been a major reason why many companies have been attracted to or become established in Derwentside. The *Derwentside Fact File* provides, in great detail, information on the availability of grants and loans, viz: Regional Development Grants; Regional Selective Assistance; Rate Relief Grants; Other Local Authority Assistance; BSC Industry Assistance; NCB/British Coal (Enterprise) Loans; European Loans; Business Improvement Services; Training; Industrial Premises; and Key Worker Housing. It is also known that DDC expended some £1,450,000 on financial incentives to companies between 1979 and 1988 (DDC Industry Committee, 1988), of which

£250,000, or 17 per cent, was lost in 'irrecoverable net debts'. The most useful picture that can be given of the scale of financial assistance relates to data for a cohort of 22 firms, defined as projects that 'matured' between April 1984 and March 1985. By 1989, 18 of these were still in operation, providing 473 jobs, of which 223 were in one company. The total amount of grant paid in relation to these 473 jobs was £1,499,000 over the period 1984/5 to 1988/9. This gives an average grant per new job of £3,169.

Although this therefore seems to be a rather cost-effective programme of assistance, in terms of cost per job created, these data relate only to one not necessarily representative cohort of firms, in a period in which firm failures were much lower than in the first half of the 1980s and central Government grants were being reduced and tied more tightly to (claimed) job creation. Furthermore, they make no allowance for post-1989 job losses in, or indeed failure of, these companies. In summary, while the data are interesting, and may be indicative, they cannot be treated as giving a definitive estimate of the level of grants and loans associated with employment generation. The claim made that there has been over £50 million of public expenditure in Derwentside between 1980 and 1988 can be related to the cost of job creation. If this was associated with, say, 1,500 new manufacturing jobs, it suggests a public expenditure cost of over £33,000 per new job.

Moreover, there is other evidence that suggests that the availability of generous grants and loans has been crucial in attracting new companies to Derwentside. In 1988 the Derwentside Industrial Development Agency presented brief sketches of nine newly attracted companies, stressing that for six (Bioprocessing; Blue Ridge; Derwent Valley Foods; Electrak; Grorud; and MultiArc) the availability of financial assistance, and the DIDA's role in assembling assistance packages was crucial in attracting them to Derwentside. Considerable emphasis was also placed upon attracting venture capital from the City of London as well as public sector grants.

Labour availability

There is no doubt that the very high rates of unemployment that prevailed in Derwentside in the 1980s, especially the early part of the decade, provided attractive labour market conditions for private sector companies. Labour availability and quality, the malleability of labour, were stressed in three of the nine company profiles (Geka, Grorud and IMP). By the end of the 1980s, 96 per cent of new companies in Derwentside were reporting no labour problems while 74 per cent were non-union (personal communication, local sources).

Environmental quality

In its post-steelworks era, as the pollutant effects of BSC's activities have disappeared, an increasing emphasis has been placed upon the quality of Derwentside's environment in the competition for investment. The district council consistently opposed applications for opencast coal-mining in Derwentside in the 1980s, on the grounds that such activities prejudice the attraction of new industries and also the development of other possible growth sectors, such as tourism. This was despite the fact that the biggest private sector employer in the district was R. and A. Young, an opencast coal-mining company.

Evidence about a new manufacturing 'local economy'

It is clearly possible to claim that new companies have had positive transformative effects on the Derwentside economy. Such claims are made particularly strongly in Carney's (1988) paper, from which the following are taken:

A substantial change has already taken place in the structure of Derwentside's local economy. There is now a wide diversity of competitive, productive and profitable new enterprises. Most of the businesses established since 1980 are indigenous, that is locally owned and controlled. (Paragraph 1.10)

A new enterprise culture is beginning to be established in Derwentside. There is a growing range of technology based businesses. A rapidly rising amount of risk capital is being invested by City of London institutions to back Derwentside businesses. (Paragraph 5.16)

Most of the new businesses established since 1980 are indigenous, that is, locally owned and controlled – a major change from previous decades when branch plants were the principal form of new manufacturing investment. (Paragraph 5.17)

Accelerated growth of a wide variety of profitable small and medium-sized businesses and a further strengthening of an enterprise culture in the district will progressively improve the employment-generating capacity of the local economy. (Paragraph 5.21)

This optimistic vision clearly owes a lot to a one-sided perception of developments elsewhere in Europe (for instance in the 'Third Italy'), where industrial development is held by some to be founded upon the emergence of a complex of thriving small- and medium-sized businesses. Derwentside apparently shares many of the positive aspects of these new cutting edges of contemporary capitalist production. The above claims are best regarded as hypotheses to be tested and some preliminary points, suggesting a rather different interpretation, can be made about employment, workforce composition, insecurity of employment and continuing dependence on State financial support.

Unemployment

It is only too clear that serious unemployment problems, with all that that implies, remain in Derwentside. The district council (1991, para. 3.7.14) point out that:

Although the District's unemployment rate has fallen considerably from the peak of 28 per cent in 1982, it remains in double figures and is above both the national and regional averages.

Net migration from the district and definitional changes affecting the unemployment count have also in their different ways helped reduce registered unemployment in Derwentside. There has been a modest, though welcome, net increase of around 1,250 jobs in manufacturing between 1981 and 1989, but this has barely scratched the surface in terms of replacing the 8,000 manufacturing jobs lost between 1978 and 1981. Moreover, by the end of the 1980s, the flow of inward investment had dried up: a survey (*Financial Times*, 27 April 1990) showed that, over the previous 12 months, Derwentside won no new inward investment projects to the north-east, nor any of the intra-regional relocations.

One symptom of this is data on the extent to which even those school-leavers who participated in large numbers in YTS schemes in Derwentside in the latter part of the 1980s subsequently failed to find employment.

Workforce composition

Against a background of labour supply greatly exceeding labour demand, companies in Derwentside have been able to exercise great selectivity in composing the sorts of workforces that best serve their interests, with profound implications for who has got the new jobs, and for the conditions on which they have got them. Not only does unemployment remain high but, within the registered unemployed, there is a hardcore of permanent long-term unemployed. Of the 3,500 registered unemployed at the start of 1990, 40 per cent had been continuously unemployed for a year or more and 25 per cent for two years or more. These comprised a mixture of predominantly middle-aged men and young people whose only experience of work since leaving school has been on Government schemes. New firms, especially inward-investing ones, have been able carefully to assemble their desired workforces, which are made up typically of young people without previous experience of work in a factory, who are very aware of the limited job prospects within Derwentside and so willing to exhibit the commitment and attitudes to the company and their work that their new employers want to see.

The significance to new employers of this vast pool of labour – in terms of enabling (if not encouraging) great selectivity in recruitment – has become particularly apparent as the first tentative signs of the limits imposed on such a strategy by local labour market conditions begin to appear. Despite the continuation of high overall levels of unemployment, companies have reported either precisely-defined skill shortages or, more revealingly, apparent deficiencies in the quality of new labour being selected from those remaining unemployed. Our interviews with employers clearly demonstrated the view that initially they had a ready supply of 'reliable' workers (low turnover and absenteeism) but were now having to take on employees who did not exhibit these characteristics.

Companies like Grorud, which were once used to considerable freedom in the local labour market, are now experiencing the first signs of a marginal change in terms of what employees ask and might reasonably expect from an employer in terms of quality of work. This underlines the extent to which new practices were easy to introduce at a time of mass unemployment, and reveals the first tentative signs of the re-emergence of bargaining power on the part of labour in Derwentside. The implications, for existing and potential new investment, and for employees, remains to be assessed, but there is no guarantee that the 1990s will not again witness rising unemployment.

Insecurity of employment

Even so, as the data on company failure rates reveal, for many of those who do find work in the new companies, precarious employment, characterised by uncertainty, becomes the order of the day. For instance, in

May 1991 one of the district's self-proclaimed success stories, Blue Ridge Care, called in the receivers. Earlier, in 1990, it had attempted a move to continuous four-shift operation. This produced little gain in productivity so it proceeded to make 44 of its newest or most troublesome employees redundant.

In solving a productivity problem, the company also neatly reshaped its workforce, further reinforcing the insecurity of local employment patterns and demonstrating – not least to its own workforce – that employers still have some considerable power in the local labour market. As has often been the case, however, this restructuring was not enough to guarantee the jobs of those who remained after it. It is also important to acknowledge that there is not only an uncertain future for those who temporarily find waged work in these new firms but also for their owners, who are often trying to make the transition from employee to self-employed and employer, at considerable personal risk and worry, fired not so much by enthusiasm for the enterprise culture as by desperation in the face of never finding work again.

Dependence on state finances

As this last point suggests, claims as to the flowering of an 'enterprise culture', centred around innovative 'high-tech' small firms in Derwentside are, at the least, premature. Indeed, they may be more prescriptive of a certain (and hotly debated) view of what is desirable rather than descriptive of conditions as they actually are in Derwentside. It may, indeed, be more accurate to suggest that what has happened in Derwentside is less the emergence of a hegemonic enterprise culture than a change in the dominant forms of state regulation there. The district is no longer dominated, as it once was, by the nationalised coal and steel industries as the main sources of wage labour there. However, their role has been taken over by other branches of the state, concerned with the management of unemployment, with a dependence upon state transfer payments rather than nationalised industry wages as the main form of monetary income for many of Derwentside's residents. As well as the 3,500 officially defined as unemployed there are over 2,000 people absorbed by various Government temporary job schemes (see Derwentside District Council, 1991, para 3.7.6). Moreover, Local Government remains the biggest employer in Derwentside: in 1986, for example, over 27 per cent of employees in employment were employed in this sector, 2,880 by Durham County Council and 1,660 by Derwentside District Council.

Figures such as these, and those on employment on schemes and on unemployment, in fact only serve to confirm the extent to which a new private sector manufacturing economy, based upon an enterprise culture, has not emerged in Derwentside. It is, therefore, perhaps dangerous to continue to believe that the solution to the district's social and economic problems lies in seeking to nourish such an enterprise culture via local economic development policies.

New business not indigenous

As the preceding paragraph points out, many of the key decisions that influence people's lives in Derwentside continue to be taken outside the district, often within the state, although it must be acknowledged that the district council is seeking to alter its mode of operation to make it much more receptive to and responsive to needs (other than those of business) expressed within the district. But, even in the more limited sphere of new businesses, and without rehearsing all the arguments as to why local control of local economies in a capitalist society is an unattainable goal based on a misspecification and misunderstanding of the issues (see Hudson and Plum, 1986), it is not at all clear as to the sense in which one can talk about the emergence of 'indigenous businesses locally owned and controlled'. Nor is it clear that most businesses established since 1980 which still survive 'are indigenous, that is, locally owned and controlled'. Many of the new businesses, especially the larger ones, result from inward investment by entrepreneurial immigrants (Caulkins, 1991). The failure rates are by far the largest amongst new start-ups which are more often locally owned and controlled.

Conclusions

In this paper, our aim has been to examine the effects of implementing a 'reindustrialisation' strategy on the manufacturing sector in Derwentside. We have sought to reveal the impacts of one chosen strategy on one de-industrialised district. We have not sought to propose alternative local strategies, although we would point out that both the absolute volume and rate of manufacturing employment growth in Easington District exceeded that in Derwentside in the 1980s, largely as a consequence of the activities of the now-defunct Peterlee Development Corporation. Indeed, we would question the extent to which local problems are amenable to local solutions alone. The key issue is the way in which local, national and supranational government policies shape the trajectory of local economic growth or decline. Indeed, one of the salutary lessons of Derwentside's experiences of reindustrialisation policies in the 1980s is that these were heavily underpinned and funded by national government involvement while being presented in popular discourse as local initiatives. Equally, it is only fair to point out that, in Derwentside itself, there is an emerging perception of the need for a broader and more broadly grounded socio-economic development strategy in the 1990s.

In this sense, conclusions at this stage are inappropriate – the point is to open up a series of questions and issues that might contribute to the debate about policy alternatives rather than produce conclusive policy answers. On the other hand, there is enough evidence here, in the form of the numbers and types of new jobs, the numbers and types of new companies, the extent to which these are dependent on state support, to suggest that the image of Derwentside as a 'low-tech', let alone a 'high-tech', version of the Third Italy is dangerously, if not deliberately, misleading. It only serves to distract attention from actual social and economic conditions there and to keep off the agenda meaningful discussions as to the sorts of policy responses that might be both possible and necessary if poverty and inequality, as well as economic growth through the private sector of industry, are seen as problems requiring attention. And in so far as there are many other deindustrialised localities both within the United Kingdom and the European Community, this debate about appropriate policy responses has a broad national and supranational relevance.

Note

This paper is a selective summary of a longer paper containing detailed statistical data and extended discussion of technical problems associated with different kinds of statistical data. The original is available as Occasional Paper no. 25 from the authors at the Department of Geography, University of Durham; it was first published in 1991.

References

Carney, J. (1988), *The Derwentside Industrial Development Programme: Background, Progress and Future Objectives*, Derwentside Industrial Development Association, Consett.
Caulkins, D. (1991), 'The unexpected entrepreneurs: small high technology firms, technology transfer, and regional development in Wales and north east

England', in M. Blim and F. Rothstein (eds), *Anthropology and the Global Factory*, Praeger.

Derwentside District Council (1988a), *Progress Review*, Consett.

Derwentside District Council (1988b), *Industrial Analysis*, Consett.

Derwentside District Council (1988c), *Opencast Coal Mining in Derwentside*, Consett.

Derwentside District Council (1991), *District-Wide Local Plan Industry Research Paper*, Consett.

Derwentside District Council Industry Committee (1988), *Monitoring of District Council Financial Incentives*, Consett.

Derwentside Industrial Development Agency (1988), *Presentation to Derwentside District Council*, Consett.

Derwentside Industrial Development Agency and Derwentside District Council, n.d. *Derwentside Fact File*, Consett.

Hudson, R. (1989), *Wrecking a Region*, Pion, London.

Hudson, R. and Plum, V. (1986), 'Deconcentration or decentralisation? Local Government and the Possibilities for Local Control of Local Economies', in M. Goldsmith and S. Villadsen (eds), *Urban Political Theory and the Management of Fiscal Stress*, Gower, Aldershot, pp.137–60.

Hudson, R. and Sadler, D. (1983), 'Anatomy of a Disaster: the Closure of Consett Steelworks', *Northern Economic Review*, 6, 2–17.

Hudson R. and Sadler, D. (1989), *The International Steel Industry: Restructuring, State Policies and Localities*, Routledge, London.

Hudson, R. Schech, S. and Krogsgaard-Hansen, L. (1991), *The New Service Economy of Derwentside District*, University of Durham Department of Geography, Occasional Publication (New Series).

Kearney, T. (1990), *Painted Red: A Social History of Consett*, 1840–1990, DCA, Delves Lane, Consett.

Robinson, F. and Sadler, D. (1984), *Consett after the Closure*, University of Durham, Department of Geography, Occasional Publication (New Series) 19.

PART III
INDUSTRIAL REGENERATION THROUGH NEW INVESTMENT BY PUBLIC AND PRIVATE SECTOR PARTNERSHIP

6

THE EASTERN RUHR DEVELOPMENT AGENCY: A PUBLIC AND PRIVATE SECTOR INITIATIVE FOR STRUCTURAL CHANGE

DR HANS ESTERMANN AND
FRANZIS ROXLAU-HENNEMANN

The importance of land in regeneration policy

Industrial and commercial land is the most important factor when it comes to securing sites and relocating businesses. This is something which has been experienced in recent decades, particularly in the Ruhr. Although infrastructure, workforce potential, research and educational facilities, residential accommodation, leisure and culture, as well as financial incentives and the region's image, have a special part to play, land sites have always been and still are of paramount importance.

Huge areas of steelworks and other iron and steel industry installations, almost as big as parts of cities themselves, have been absorbed into the centres of Duisburg, Essen, Bochum and Dortmund. Relatively small mines cover the Ruhr and have characterised the structure of settlements in the region. Settlements grew up around the blast furnaces and pits, growing together into towns within a century.

However, coal and steel distinguished not only the settlement structure, but the economy too. Their dominant position was expressed in the variety of supply and sales connections, as well as in the communities' dependence on major companies. A consequence of this was that significant sections of the land market were also controlled by the coal and steel companies. Above all, land was stockpiled for future company developments or, in the case of mining, for future exploitation. Here are two examples. In 1954 the town of Bottrop had 4,200 hectares of land; 2,173 hectares of this belonged to mining, 806 hectares to the public authorities, 918 hectares to private owners, 77 hectares to housing construction companies and 40 hectares to churches. In 1958, Castrop-Rauxel had 4,400 hectares of commercial land, of which 1,222 hectares belonged to mining and 216 hectares to other industrial concerns. After Ruhrkohle AG, the major national coal firm was founded, it only took over industrial land, which means that Ruhrkohle's share of the commercial land is no longer as great.

However, it is important not to limit ourselves to quantitative observations. Most of the land owned by industry has planning approval. In Dortmund, for instance, very little new industrial land is owned by the city, while the steel industry has large reserves. Planning restrictions are also the most important aspect when considering development opportunities. The settlement structure and impact of the area's infrastructure, industry and commerce mean that there is virtually no potential for new sites. In addition, new sites would be expensive to develop, whereas pits and steelworks have an adequate infrastructure.

On the other hand, new space for business is limited for ecological reasons. No more land can be released for business use except by reclaiming derelict industrial land; otherwise a shortage of land will limit the potential of the regional economy. Under market conditions the cost of reclaiming derelict land had to be included in the price of the land. But as a rule there is no tradition of a 'land market'.

From a historical point of view, it was not in the interests of coal and steel concerns for other industries to settle in the Ruhr. They did not want competition on the labour market and, in view of their economic situation, they had every reason to pursue their stockpiling policy intensively. When this attitude changed towards the late 1970s, there was no market in the Ruhr, at least not for land containing waste products, where steelworks and coking plants had previously stood. Nobody knew what sort of expenditure would be involved in reclaiming the land and, above all, there was the risk of liability for subsequently discovered waste products. Sensitivity to such questions was also growing at this time, as a result of cases in which buildings had been constructed on waste products.

The fate of land after pit closure

During its migration northwards within the Ruhr, mining has left behind a particularly large amount of land. The initial obstacle to its redeployment has been the fact that it is subject to the Mines Supervisory Board until the operation has been shut down, with the demolition of the typical mining installations and securing of the land. Arriving at solutions to this which would be both cost-effective for the mining company and beneficial to the community's urban development policy requires intensive and complex co-ordination, co-operation and financing processes. These have not existed in the past. Mining companies typically closed the pit, sometimes renting out parts of the land or putting it to some other use. Only afterwards has the community been able to start its planning process and has generally failed to find a starting point for financing a redeployment scheme.

Hence we have the sort of situation where a former pit site, which closed down over 40 years ago in a prime location in the south of Dortmund, is waiting for a satisfactory use. It is currently being used mainly by scrapyards and minor repair shops. Similar cases are common throughout the Ruhr, although the sites are often not owned by RAG itself, but by its predecessors.

Despite numerous problems, it has been possible to convert a series of former pit sites for industry and commerce. The communities have often bought land, prepared and developed it, and then resold it. Parks and public facilities have grown up there too.

Intervention by regional government

The first breakthrough came with the intervention of North-Rhine Westphalia's government, as part of its Ruhr Action Programme between 1980 and 1985. During this time, 6.9 billion deutschmarks were to be invested in the Ruhr, in order to give the crucial kick-start to structural change. Of this, 100,000,000 deuschmarks a year were to be spent on acquiring, developing and marketing disused sites. Responsibility for performing this task went to the Land Development Company (LEG), which works in close collaboration with the Land government and communities concerned.

Since then, over 20 million square metres have been purchased. However, only approximately four million have been resold: 1.7 million for businesses, 2.2 million for parks and leisure areas and 150,000 for housing. In this case, the LEG has gone beyond simply developing and marketing the land and has, for instance, participated in the newly created business zones. These include a Centre for Ecological Planning and Building in Hamm, promotional and technology centres, a countryside park in Duisburg and house-building projects.

In addition to the Ruhr property fund, there has also been a Land fund since 1984, with which the LEG is active throughout North-Rhine

Westphalia. Its specific objective is not exclusively to create industrial and business areas: it has a more far-reaching involvement in urban development especially and the relocation and upgrading of housing and recreational space in the Ruhr. The property fund has therefore become a major development sponsor for a number of communities. The protracted nature of the processes and the number of areas involved has stretched the property fund to its very limits. There is hardly any scope for further projects, particularly complex expenditure on the acquisition, rehabilitation, development and marketing of former pits.

The property fund is an instrument used exclusively by the LEG. The Ruhrkohle AG and other property users played no part in the process. There is, however, a great deal of interest in making property owners accept joint responsibility for land reclamation. They should take an active part in developing and marketing it and also be jointly liable for the financial risks. The first attempts at this were made in the context of RECHAR projects, in which subsidies of up to 80 per cent were given from Land and EC funds for the development and preparation of ex-mining land. However, these subsidies can only be given to private concerns under very restrictive conditions. For example, it must be guaranteed that the public authorities have a decisive influence over use of the developed land for 25 years and the use of the land has to meet the economic structural aims of the local authority.

Based on close collaboration between the Land government, Ruhrkohle and the communities, organisational and financing models were developed for this. The industrial development areas of 'Prosper' in Gladbeck-Brauk and 'Minister Stein' in Dortmund Eving have emerged to date. These are both part of the Emscher-Park International Building Exhibition.

In 1992 a project company was founded in Dortmund by the city and Ruhrkohle, both having a 50 per cent stake. It was contractually agreed, however, that the city should have the majority vote in all cases involving the use of promotional funds. Roughly 30 million DM are to be spent on ten hectares of industrial land. About 20 million DM of this will be covered by public subsidies. Ruhrkohle puts up about 8 million DM, while the city meets the local government shares of the development work and parts of the equity share of the schemes being promoted. Subsequent proceeds of up to 10 million DM are expected from the property, so that the mine's equity shares would be covered.

This model requires a high level of expenditure by Ruhrkohle, which it will hardly be prepared to meet in future. Consequently, it is not really possible to apply this model directly to other projects, although the principle of joint responsibility on the part of the property owners and public authorities is probably accepted. It also seems to be the only feasible way to mobilise unused industrial land, outside the property fund. However, the organisational expenditure behind this must not be underestimated. It affects all those involved. The Land government is represented by several departments; it is impossible for smaller communities, in particular, to process and implement the organisational and financial planning and Ruhrkohle, with its subsidiary Montan-Grundstücksgesellschaft (MGG) which is responsible for this, will not provide further input to the extent it did in Gladbeck and Dortmund.

The creation of the Eastern Ruhr Development Agency

In 1992 the Land government proposed to the regions, in the context of the regionalisation of structural policy and a special support programme for coal, that the initiation, co-ordination and also joint marketing of land reclamation projects should be transferred to a regional agency. Regional bodies, chambers of commerce, business organisations and major property owners were to work together to form this agency for the Eastern Ruhr. This proposal, which came with a promise of 10 million DM of start-up finance, was seized upon in the region. Within eight months, a limited company had been set up with the following members:

- the city of Dortmund, the district of Unna, the city of Hamm, the city of Ahlen;
- chambers of industry and commerce and Handicraft, the Retail Traders' Association; and
- Hoesch, Krupp, RAG, Harpener, LEG.

Its most important function is to mobilise significant areas of disused land in the Eastern Ruhr. The Eastern Ruhr Development Agency (EWA):

- assumes responsibility for preparing projects, financing expert reports on risk assessment, planning concepts, and for rehabilitation concepts and financing;
- co-ordinates these plans, since reclamation needs to take account of potential use;
- prepares the project organisation, in the form of a project company, for instance;
- co-ordinates the various projects within the region;
- monitors the project company, whose functions lie primarily in preparation and development; and

- markets land and undertakes joint site marketing.

The EWA has six projects to begin with, which it will focus on over the coming years. These are the former mining areas of:

- Achenbach (Lünen) 44 hectares;
- Radbod (Hamm) 40 hectares;
- Gneisenau (Dortmund) 52 hectares;
- Königsborn (Unna) 9 hectares;
- Werne (Werne) 20 hectares; and
- Westphalia (Ahlen) 40 hectares.

The estimated amount of finance is approximately 150 million DM, which will be raised over the next five to six years.

In addition, individual functions are to be assumed for other projects and it is conceivable that, in the event of further shutdowns or partial shutdowns within the region, reclamation will be started quickly with the help of the EWA. The projects are chosen from 55 possible sites in the region, which it will be possible to develop into additional industrial land in future. Not all of these involve disused land and not all of them are beset with problems which would require the use of an agency. There is no doubt that in the short term it will not be possible to implement the six projects chosen. It will therefore be necessary for communities to be able to market other industrial land in the short term. In some cases this can only happen in free space, and land planning will only allow this on the condition that there will be sufficient reactivated land available in the foreseeable future, so that any further use of free space is avoided.

The problem of finance

EWA costs which are not covered by subsidies from the Land are divided between the public and private sector partners. Nevertheless, these currently amount to five hundred thousand DM per year. This sum is small, however, compared with what is required to develop the land. This requires particularly intensive co-operation between the agency partners. In view of depleted local government funds and the catastrophic economic situation in the mining industry, there is a risk of projects collapsing due the relatively low equity shares; full state backing cannot be achieved, however.

The financing models will be crucially dependent on how the problem of intermediate financing can be solved. Of prime importance is to be able to guarantee the sale of the land once it has been reclaimed and, in view of the current economic situation, this is difficult. However, it is absolutely essential for projects to be started now, so that sufficient high-quality land can also be made available for business and industry in future too, in the face of regional competition.

7

THE CAMPAIGN FOR THE FUTURE OF COAL-MINING AREAS: A LOCAL GOVERNMENT PRESSURE GROUP FOR GERMANY'S COAL-MINING AREAS IN EUROPE

WOLFGANG STEINGRÄBER

This chapter is divided into four sections: firstly, the emergence of ZAK and EURACOM; secondly, the functions and aims of ZAK; thirdly, the practical work undertaken by ZAK; and fourthly, a five-point summary of ZAK.

The Campaign for the Future of Coal-mining Areas has both national and European roots

In September 1987, at the instigation of the town of Hamm, a meeting took place between the towns of Ahlen, Bergkamen, Hamm, Kamen, Lüben and Werne. This mini-Coal Round passed the 'Hamm declaration', which called for politicians at Land, Federal government and EC level to guarantee the long-term security of German hard coal. I quote:

In the past, mining has used its strengths to supply energy to a society weakened by need, for the sake of solidarity, to enable it to survive difficult times and to pave the way to prosperity. Now mining finds itself weakened and threatened by external influences; it needs the solidarity of a strong society, because without it a whole region is condemned to death.

Meetings then followed at Land, Federal and EC level, aimed at giving weight to these demands. In November 1988, a follow-up conference was able to report that the Hamm declaration had exerted a positive influence on the coal policy resolutions passed by the Federal government in 1987, although this had not allayed fears in the region. Only one thing was clear – hard coal mining was not a museum industry and a viable, efficient mining industry was still needed in order to guarantee the domestic energy supply.

There were allies too. Under the slogan 'The area needs structural change not structural collapse', the round of talks involving Lord Mayors and Lady Mayoresses from the Ruhr had as its central theme the maintenance of a long-term perspective for German hard coal. The Round fought for the view that the economic restructuring of the Ruhr and new growth and employment incentives could only come from hard coal mining in its industrial framework, not from its ruins.

The activities reached a climax on 29 August 1989 at a major forum held in the Maximilian Hall in Hamm under the slogan 'Into the future with coal'. The Prime Minister of North-Rhine Westphalia, the Federal Republic of Germany's employment minister, the heads of steel, energy and mining companies and the Chairman of the Mine Workers' Union spoke out in favour of keeping German hard coal in front of over a thousand participants from North-Rhine Westphalia's mining communities.

The campaigns conducted between the Hamm Coal Round in 1987 and the Coal Conference in 1989 led to a forum in Herten in 1990, at which the idea of a jointly organised delegation representing the interests of all Germany's mining regions took on a concrete form. On 23 November 1990 42 towns, communities and districts from North-Rhine Westphalia, 18 towns from the Saarland, the National German Hard Coal Mining Association and the Mining and Energy Union came together in Hamm to form the Campaign for the Future of Coal-mining Areas (ZAK). ZAK has no party political affiliations and speaks with one voice for mining and its regions. Its membership has now grown to 70. The new Federal Lands, and with them, for the first time, a brown coal region, are represented by members from Saxony and Saxony-Anhalt.

Efforts to unite coal-mining areas at European level ran parallel to the national amalgamation of coal-mining areas. Europe's time came in 1988. Over two days in September the first conference of the European mining communities took place in Strasbourg. Members of the European Parliament had been invited to attend, including Ludivina Garcia Arias from Spain, Norman West from Great Britain, Lambert Roux from

Belgium, Jean-Marie Alexandre from France and Hans Peter from Germany. They had all decided, due to the crisis situation which had persisted for years in the European Community's mining regions and at the request of members of their national constituencies, to discuss matters of energy security, structural change, eligibility and economic recovery and, along with local government, to demand a Community Programme for the affected areas.

The conference resulted in the formation of a working party, ACOM, which called upon the representatives from Germany, France, Spain and Belgium to found a national lobby along British lines to represent their mining communities. ACOM assumed the task of promoting coherence between the European mining regions, enforcing the claim for an EC programme for coal-mining districts and preparing the annual conference of Europe's mining regions. The first rewards were reaped at the second conference of all European mining regions in November 1989. The working party of European mining regions, ACOM, had succeeded in persuading the EC Commission to start RECHAR, a development programme specially designed for coal-mining areas.

Eighteen months later, at the third conference of European mining regions in March 1991, the time had come for representatives from the five European countries – Great Britain, Germany, Belgium, Spain and France – to come together officially as EURACOM and establish its constitution. The function of the European Campaign for Mining Regions is to procure funds for economic and social recovery and environmental rehabilitation, to implement energy policy measures aimed at promoting the use of coal from the European Community, to promote exchanges of information and experience within the coal regions, to advise members and represent their interests within European Community institutions and to work closely with the inter-party working groups of European Parliament members from mining areas.

The president of EURACOM is Mrs Ludivina Garcia Arias and her deputies are Professor Sabine Zech and Jean-Pierre Kucheida. EURACOM's members are: Great Britain, with 97 towns and communities; France, with over 180; Belgium, with over 10; Germany, with 70; Spain, with over 100; the Netherlands, with 10; and Portugal, with 2. There is now officially a joint delegation representing the interests of all European mining regions, which can be regarded as a regional lobby based on increased regional co-operation. The European mining region movement has finally taken root and has in the meantime grown into a mighty tree. One of the strong branches of this movement is the Campaign for the Future of Coal Mining Areas (ZAK).

What are the functions and aims of ZAK?

Its function and aim is to improve living conditions in coal-mining areas. In terms of energy policy, we want coal to remain the primary energy reserve and, in terms of structural policy, we want to forge ahead with the structural change which is already underway. We want to maintain links with the regions which have already completed their structural change or are well on the way to doing so. We want to change structures, but not to destroy them. And we want this process to be socially tolerable, in other words, not to involve job losses.

Structural change is the magic concept on which the performance of our economy finally rests. It can be seen in the constantly changing quantity and quality of jobs. Jobs are being cut, on the one hand, while on the other, new ones are emerging. If this does not happen, there is a risk of structural collapse. We can only speak of structural change when job cuts and job creation go hand in hand. We are therefore convinced that conditions in our regions can only be improved by promoting structural change, while at the same time maintaining a modern coal and steel industry. Structural change only makes sense if it is based on a viable, efficient mining industry, not on its ruins. We are, and remain, an industrial society and must go back to our roots to build up new developments. The slogan is 'Yes to structural change. No to structural collapse.'

As the Campaign for the Future of Coal-mining Areas, we are the model case for regional thinking. We do what local government alone cannot do. We work as an agency for structural change, offering services to our members. We have three main areas of activity. As the delegation representing the interests of mining regions to Federal government, the Land, the EC and companies, we are the binding element internally and a powerful instrument externally. We lobby through political action, discussions, resolutions and input and launch initiatives for schemes, projects and programmes. As an information exchange, we are the contact point for queries and ideas. We are the turntable and transfer point for plans, schemes and opinions. We carry on foreign policy for mining regions, canvas for mining to be accepted within the population and clarify political decisions. We spread news and are involved in public relations work through our information service, press releases, press conferences, interviews, exhibitions, forums and the organisation of visit programmes. As a forum for ideas, we create contacts, offer an exchange of experience, organise co-operation and act as a moderator when it comes to action on coal and structural policy. We want to be the workshop for the future development of our regions in platform debates, conferences and seminars.

What are our objectives and successes?

From a European point of view, our greatest success has been in bringing together the Belgian, British, French, Spanish, Portuguese, Dutch and German mining regions to form a joint organisation, EURACOM. Regionalisation, as an element of decentralised politics, is becoming increasingly important, not only at the European level but at the Federal level too. The need for collaboration throughout Europe and the Federal Republic is undisputed. As an organisation, EURACOM is effective simply because it exists. Without EURACOM, the RECHAR programme for the economic restructuring of coal-mining areas would not have been started.

The seriousness with which the EC Commission also views the amalgamation of Europe's mining regions is reflected in the fact that the Commission is financing a seminar programme for an exchange of experience between member states. Seminars have been held in Great Britain, Belgium, Germany, Spain and France. These have provided an opportunity to explain how individual countries are handling the funds received from RECHAR, how they are combating unemployment, cleaning up the environment, promoting new economic activities and controlling mining damage. All EURACOM's member organisations agree that the seminars are not only a practical way of exchanging information on common problems, but that they also strengthen the ties between our members and interest the EC Commission in the special problems of mining regions.

On a national scale too, the coming together of the Campaign for the Future of Coal-mining Areas must be assessed as a major success in itself. Since our formation in Hamm in November 1990, 10 more towns have joined us. We represent 70 towns, communities and districts with eight-and-a-half million inhabitants. The addition of towns from the new Federal Lands means that we represent not only hard coal, but brown coal regions too. What is particularly important, however, is that ZAK's members include not only local government, but companies and trade unions too. This is an example of a public/private sector partnership within a pressure group and serves to promote joint projects between local government and companies. As a local government public service enterprise for structural change, ZAK has organised seminars and events and brought the players on the political and economic stage to one table.

If the EC, Federal government and Land today announced special programmes for structural measures in coal-mining regions, the Campaign for the Future of Coal-mining Areas would have played a part in this. Whether it is the RECHAR programme or the joint initiative between Federal government and the Lands to improve the regional economic structure, whether it is the 1991 Coal Round with Coal Concept 2005 or the exemplary working framework for North-Rhine Westphalia's coal areas, the coming together of the coal-mining regions to form the Campaign for the Future of Coal-mining Areas has played its part as the pioneer of structural change and will continue in future to represent the interests of coal-mining regions independently and impartially, going beyond party boundaries.

A five-point summary of ZAK:

- As an independent association, the Campaign for the Future of Coal-mining Areas is the local government delegation representing the interests of Germany's coal-mining areas.
- The Campaign for the Future of Coal-mining Areas provides a model for regional thinking from below, both at national and European level.
- The Campaign for the Future of Coal-mining Areas is an example of a voluntary public/private sector partnership.
- The Campaign for the Future of Coal-mining Areas is a local government public service enterprise dealing in matters of structural change.
- The Campaign for the Future of Coal-mining Areas is the protagonist of a long-term perspective in energy policy, based on the use of domestic energy sources.

8

BRITISH COAL ENTERPRISE

DAVID PICKERING

Origins of the company

Between March 1980 and 1985, 50 collieries closed within the British Coal Corporation. 1984/85 was also the time of the major strike in the United Kingdom Coal Industry. It was also the time that British Coal Enterprise was formed: during the winter of 1984/85 in the midst of the strike. In the eight years since the formation of British Coal Enterprise the number of collieries has reduced by 119, proving an ongoing need for the services offered by British Coal Enterprise. The aim of the organisation is to help create jobs and stimulate economic regeneration in the coalfield areas of the United Kingdom. The important point to note is that British Coal Enterprise only helps to create job opportunities by working with other people and agencies. They do not claim to have done everything single-handed. British Coal Enterprise has played an essential part in British Coal Corporation's strategy to achieve radical restructuring with minimal industrial disruption.

British Coal Enterprise is a wholly owned subsidiary of British Coal Corporation, which is itself a state-owned nationalised company. British Coal Enterprise's revenue costs of operation are borne by British Coal Corporation and its capital fund has been progressively increased to £60 million. Standard U.K. Treasury rates of interest are paid on this loan fund, which are currently just less than 8 per cent. Although British Coal Enterprise's outplacement and associated training services are available free of charge to ex-British Coal employees, they are also offered commercially to other companies within, and also outside of, the coalfield areas.

The organisation

British Coal Enterprise operates on a regional basis, within the boundaries corresponding to the seven BCE Coalfield regions of the United Kingdom – Scotland, Northern, Yorkshire, Midlands, Western, Wales and Kent. Each region is served by a regional manager and supporting staff who provide the full range of British Coal Enterprise's services in that region. Some members of this regional team are seconded from banks and accounting firms, which gives a blend of expertise throughout the company.

The fields of activity for British Coal Enterprise are:

- business funding;
- support for enterprise agencies;
- managed workspace and small business development;
- outplacement services; and
- new initiatives, which includes consultancy work in Europe, training products and skills training centres.

Business funding

Business funding is the 'core' activity of the organisation. For several years British Coal Enterprise concentrated on the loans business. Loans and/or equity funding are provided to companies setting up, expanding their employment base or locating into a coal mining area and creating new jobs. Loan clients come to the organisation either directly, or are referred from banks, local enterprise agencies or professionals such as accountants and solicitors. Loans are normally between a thousand and a million pounds for a term of up to five years, with a provision for a capital repayment holiday of 12 months. The loans are repayable at a fixed competitive rate of interest, about 2 per cent above base rates. This is approximately equal to British Coal Enterprise's borrowing rate from the U.K. Treasury. It is the policy of British Coal Enterprise to break even on their interest account. Security arrangements on loans are less stringent than the banks in that security is only required on loans over £25,000 – a higher threshold than banks normally adopt.

Equity is provided between £25,000 and £250,000 in the form of preference or ordinary shares and may be combined with a loan. Policy has always been to withdraw after five to seven years with shares having a net 10 per cent fixed coupon. Analysis of this type of investment indicates an internal rate of return of just over 20 per cent, which is considerably less than the normal venture capitalist would demand. British Coal Enterprise fills a gap in the equity funding market in that most venture capital companies will not consider equity investments of less than £250,000.

There are certain qualifying criteria which must be met in order to obtain funding: new permanent jobs must be created through the life of the loan which is normally assessed over a three-year period; the business must not displace jobs in other businesses; the business must be located within the coalfield areas; funding is only available for up to 25 per cent of the total funding package or £5,000 per new job created, whichever is the lesser; start-ups, expansions and inward investments by a business all qualify; and the business must be considered viable by British Coal Enterprise, after the submission of a detailed business plan by the applicant, which must include a cash flow statement and evidence of market research.

Although the businesses which British Coal Enterprise assists cover the whole spectrum of activities, there is a strong preference for encouraging and supporting manufacturing industries. In fact, 60 per cent of the businesses assisted are in manufacturing and 40 per cent in service industries.

Although many loans are largely unsecured, this activity has proved to be successful. To date, British Coal Enterprise has received repayments totalling £31 million, with approximately £600,000 received each month. Nine per cent of funding has been written off as bad debts, with a further 22 per cent of companies in repayment difficulties. Where a company encounters difficulties, rather than look to protect their own security, British Coal Enterprise involve themselves directly with the management of that company to see what steps can be taken to help them. Inevitably some companies will go into liquidation, but it is thought that the assistance offered through our Business Support Executives is a key factor in the survival of many businesses.

By March of 1993 the following had been achieved:

- 3,775 projects had received funding throughout the coalfields of the United Kingdom;
- 33,499 jobs were in place from the projects supported;
- 41,647 jobs would eventually be created as a result of this funding reaching maturity;
- total investment in the 3,775 projects had exceeded £589 million including £75 million from British Coal Enterprise;
- the amount of investment per job was £2,078; and
- taking into account all administrative costs and costs of bad debts and company failures, the cost per job for 1992/93 was £2,300.

Support for enterprise agencies

Enterprise agencies are typically independent local organisations whose prime function is to provide business planning and other advice to start-up businesses. British Coal Enterprise has encouraged the setting up of local enterprise agencies – often being the major provider of funding, while continuing to provide management advice and stimulation to the Enterprise Agency movement. Alongside this, British Coal Enterprise works with local government economic development units, training and enterprise councils (TECs) and other regional enterprise bodies. Six million pounds has, to date, been granted as aid to enterprise agencies throughout the United Kingdom. In the last financial year, British Coal Enterprise assisted some 118 enterprise agencies with financial support of just over half a million pounds. British Coal Enterprise is a key player in local networks. Breakfast sessions and early evening seminars are hosted regularly, not only to describe the work carried out by British Coal Enterprise, but also to bring together the various enterprise bodies, encouraging them to work more closely with each other.

Managed workspace

British Coal Enterprise's other main area of activity is the provision of managed workspace for small businesses. This accommodation varies from 200 to 5,000 square feet in area, with units averaging 1500 square feet. One of the major features of all sites are 'easy in, easy out' leasing terms which do not tie the tenant to a long-term lease. This means that the tenant has an opportunity to try out their business idea, and if it does not work, or they want to expand to larger premises, they can leave freely with usually less than one month's notice. There is no other severance charge. On the larger sites administrative facilities are made available, which include typing, facsimile and reproduction services, carried out by a receptionist.

Originally managed workspace was provided through conversions of former British Coal buildings. But, as demand has increased, workspace has sometimes needed to be built specifically for this purpose, on land which may not be owned by British Coal but rather in joint development with local authorities. A key feature of such schemes is that they attract European grants of 40 per cent. With this assistance, projects generally produce an internal rate of return of between 10 per cent and 11 per cent.

British Coal Enterprise operates in a commercial manner but clearly could not survive if they had to borrow money in the open market.

At March 1993, 24 sites were operated by British Coal Enterprise with 500 units, totalling 760,000 square metres in area. A large capital programme was in place (with £17 million to be spent in the next 12 months) and by March 1994 British Coal Enterprise plans to operate 41 sites with 790 units, totalling 1,160,000 square feet of space. As part of the same initiative, British Coal Enterprise has also given grant aid to other organisations to assist them in providing managed workspace. This investment in workspace has created 12,000 jobs to March 1993, at an average capital investment of £2,130 per job created, with a revenue cost (including all capital charges) of £700 per job created.

Outplacement services

In 1987, British Coal Enterprise took over the job and career change scheme (JACCS) from British Coal Corporation and in 1988 it re-launched the scheme. Outplacement services are available to all redundant employees of British Coal with a personal career counsellor and unlimited counselling provided for every individual. Additional services offered include self-marketing and interview technique training, and help in c.v. preparation. All clients receive a job seekers pack or career development pack (containing home study work books and training videos). Finance may be made available for clients who wish to train further in order to improve their career prospects, but this must be geared to employment, and in accordance with employment market demands.

For those considering self-employment, a one-day course is run explaining the rewards and risks of entering into self-employment and helping clients assess whether self-employment is the right option for them. A three-day intensive course then prepares clients to produce their business plan, before they are referred on to outside bodies such as local enterprise agencies or TECs for specific advice. Where individuals are interested in retiring, pre-retirement advice is offered on subjects such as how to invest their money wisely. This is carried out independently through a national arrangement with two companies.

The outplacement scheme has been so successful that British Coal Enterprise has been asked to provide similar assistance to many outside companies. This external work is carried out:

a) to gain wider experience and so enrich the work carried out for British Coal; and

b)   to help with the costs of the British Coal operation and to assist with BCE's investment in coalfield regions.

As well as the 180 career counsellors, administrative services are provided for the typing of c.v.s, letters of application and so on to aid clients moving back into the workplace, along with telemarketing teams who actively search for unadvertised vacancies to give clients an advantage in the market-place. This gives a total of approximately 350 people working on the JACCS scheme today. All of the services housed under the JACCS scheme are provided through 11 New Career Centres, set up on a regional basis for management and clerical staff, and about 30 Job Shops set up at or near each closing colliery, for industrial staff.

From these outplacement activities, British Coal Enterprise has found employment opportunities for over 34,000 people. The current cost of the JACCS operation (including retraining, where appropriate) is £1,800 per placement. The current success rate for placing redundant British Coal employees is 87 per cent. The present case load is approximately 11,000–12,000 people at any one time.

New initiatives

In the last year, British Coal Enterprise has given assistance to 95 outside organisations, providing the services and products which have been developed to help British Coal employees – including job seeker and career development packs.

Two of the products which British Coal Enterprise has developed specifically for the commercial market are the *Breaking the news* and *No simple choices* self-help packs, for people managing redundancy programmes. Other products sold commercially include *Making and taking career changes*, on sale at book stores throughout the U.K., and the *Training needs guide*, which helps small to medium-sized enterprises assess their training needs so that 'in house' training and recruitment can be geared to the needs of the business. British Coal Enterprise also provides consultancy services in Europe passing on all its areas of expertise. The current portfolio of overseas work includes the Czechoslovakian Republic, Hungary, Poland, Slovenia and Bulgaria.

Based on the success of skills training centres in Yorkshire, it is planned to open more centres throughout the U.K. The centres are aimed at improving British Coal employees' job prospects by working in close liaison with training and enterprise councils, training organisations and, in particular, employer-led organisations to provide the training that employers want, to the standards that employers require. All of the courses carried out are accredited with certificates issued on completion.

Summary

Through a diverse range of activities British Coal Enterprise has assisted in economic regeneration throughout the United Kingdom. There have been over 3,300 businesses, requiring total funding of £544 million. British Coal Enterprise has provided £70 million, or 13 per cent, of this. Funding activities have created 41,000 job opportunities, while managed workspace projects have created 12,000 jobs, and through JACCS and the New Initiatives, 34,000 jobs have been created. It can be seen, therefore, that in its eight years of existence to March 1993, British Coal Enterprise has assisted in the creation of 87,000 job opportunities.

PART IV

FOSTERING ENTREPRENEURSHIP THROUGH ECONOMIC AND PSYCHOLOGICAL INCENTIVES

9

THE PROMOTION OF BUSINESS ACTIVITIES IN THE CONTEXT OF STRUCTURAL POLICY IN NORTH-RHINE WESTPHALIA

WULF NOLL

Many individuals and institutions are involved in the processes of structural change. We, in the Federal Republic of Germany, have the added peculiarity that our administrative system is heavily structured. All this makes structural change an exceptionally complex process, the many facets of which cannot be described in a few minutes. To at least convey a general idea of what has happened to the structural policy in our country, I would like to proceed chronologically, taking my bearings from the various programmes which the government of North-Rhine Westphalia has launched since the 1960s.

The post-1958 crisis

While the post-war era had initially been entirely devoted to reconstruction, 1958 marked the start of the mining industry's demise. This signified a crisis in a sector of the economy which was at that time one of its mainstays. The industry was mainly concentrated in the Ruhr, where it still employed roughly 15 per cent of the total workforce and about the same number again in associated sectors of the economy. The consequence of these pit closures was that from 1961 the number of jobs in the Ruhr dropped and from 1965 so did the number of inhabitants.

The Land government began at a relatively early stage to tackle what were for it, too, new circumstances. As early as 1964, several government reports made it clear that what was needed to create new jobs and income potential was an intensified and coordinated infrastructure development programme. But by 1968 this had been little realised. One exception was the foundation of the Ruhr's first university. The Land's Christian Democratic government was finally toppled as a result of the general political situation, but also due to this vacillation on the question of structural policy.

First interventions in the late 1960s and early 1970s

The new Social-Liberal government in North-Rhine Westphalia started its work with the 'Ruhr Development Programme 1968–1973'. This programme already had a basic structure, which was to be essentially repeated in later programmes too. This involved:

- retraining mining personnel and providing for their social security;
- promoting private sector investment;
- improving the environmental situation;
- advertising and providing information on the Ruhr; and, above all,
- expanding and modernising the infrastructure in the areas of housing, traffic, education and recreation.

The 'North-Rhine Westphalia Programme 1975' followed shortly after and had largely the same content.

In the meantime, the economic situation had deteriorated. Following the first oil crisis, the steel industry had become the second industry to fall into decline after the mid-1970s. The numbers of unemployed now rose sharply. In 1970 unemployment in the Ruhr had been running at 2 per cent. By 1974 it was already 4 per cent and 10 years later it had even reached 15 per cent.

New interventions: technology transfer in the 1980s

The next programme aimed at promoting structural change in North-Rhine Westphalia was called the 'Ruhr Action Programme 1980–1984' and was set up in 1979. Once again, it adopted the main focal points of its predecessors, but two points were added. Firstly, special attention was now being paid to the creation of training places and help for the unemployed.

The second aspect, however, was of particular significance to our subject matter. University research had hitherto played a part in previous programmes but the main emphasis had been on policy-oriented research, with a clear bias towards the social sciences, the area of peace studies being an example. Now, economically valuable research came to the fore. Funds were also available for private research in companies and establishments involved in the transfer of technical know-how from research institutions to companies were founded.

This was a qualitative leap. In my view it stemmed from the disappointment over the fact that the situation had tended to deteriorate rather than improve, despite what had now been 10 years of endeavour to achieve structural change. Until then, the programmes had been geared towards improving the so-called basic conditions, such as infrastructure. And these had quite evidently improved. But the investment opportunities being generated were obviously not being recognised and exploited by the economy. Why? What was standing in the way of businesses?

One of the obstructing factors was identified at the time. It was recognised that investment in new economic fields of activity also required, among other things, knowledge of innovative potential which existing companies rarely had. Ways had to be found, therefore, to transfer this knowledge to companies. This policy of encouraging technology transfer has proved viable and vital to the success of structural policy in North-Rhine Westphalia. Today there are 48 technology centres and 23 technology agencies in North-Rhine Westphalia, which give advice on key points of particular subject areas, as well as a series of technology initiatives and networks. They are financed by the Economic Technology Programme, which is available as a promotional programme for such institutions and to aid research within companies. In 1992, grants of 87 million DM were promised from this for infrastructure schemes – company projects received 29 million DM of backing.

1987 onwards: the decentralisation of regional policy

The next qualitative leap in North-Rhine Westphalia's structural policy came in 1987. As mentioned, the crisis in the steel industry began in the mid-1970s. Until then, such crises had been nothing out of the ordinary in companies. So, to begin with, the motto was 'Hold on, the recovery is bound to come!' But in 1982 the so-called 'steel moderators', who had been commissioned by the steel companies, made it clear that production capacity would have to be reduced. This marked the start of a wave of closures, which particularly affected the cities of Hattingen, Oberhausen and finally Duisburg. This was the scene of violent protests, which finally climaxed in the closure of the Rhine bridge at Duisburg-Rheinhausen. Even the Federal Chancellor himself felt it necessary to hold a conference in the Ruhr. The Land government reacted with, among other things, new programmes, including the Future Initiative for Coal and Steel Regions (ZIM), followed by the Land Investment Programme (LIP) and finally the Future Initiative for North-Rhine Westphalia (ZIN).

What was new about these programmes was that they no longer had the rigid, Land-based structure of their predecessors. Instead, the Land was divided into 15 regions and the most important groups and individuals in these regions were left to decide what was important for their future development. This resulted in lists of proposed projects which could then be implemented, provided there were sufficient funds available. Experience of this proved positive. As a result, in 1990 the regions were asked to further pursue this new approach of decentralisation, co-operation and co-ordination. The initial aim was to formulate regional development strategies. These were to establish and identify the most significant development opportunities and bottlenecks.

In 1991 the first regional development concepts could be used for the next programme. This was called the Action Framework for Coal-mining Areas and came in response to new pit closures in North-Rhine Westphalia. What it basically pointed out was simply that regional development strategies should be used when elaborating the programme. It also contained a few procedural rules, focal points in terms of content and, of course, details of the programme's financing. Today we are in a transformation phase. I believe that the experience to date has been positive. The formulation will prove to be correct and workable.

The problem of private companies

Nevertheless, a substantial deficit is now becoming evident. The Land government has left the regions to decide how the regional boards, the so-called regional conferences, should be composed. As a rule, they include representatives from the chambers of industry and commerce, trade unions, resident universities and, in some cases, environmental and social groups, as well as local government. However, up to now the companies themselves have rarely been prepared to participate directly in these conferences. But if we are to discover their specific needs and be able to focus our assistance on these more selectively than in the past, we need information from these companies. We must develop solutions together and also work as a public/private sector partnership. The state is reaching the limits of what it can do without a more positive set of attitudes from private companies.

Today it is safe to assume that the thinking and action prevailing in many companies in the Ruhr is cost- rather than profit-oriented. If these companies find themselves in a crisis, they do not try to get out of their difficulties by developing new products or tapping into new markets. Instead, they try to reduce costs. However, the scope for this is becoming increasingly limited in the face of international competition. Because their main focus of attention is fixed costs, it is not unusual for important functions such as research and development, marketing or personnel planning – the production-oriented services, in other words – to suffer staff cuts, thereby limiting the company's own future prospects still further. Such companies are now outside the reach of our technology transfer instruments too.

Attitude problems also beset the major companies within our region, of which there is an above-average number. They have never actually learnt how to sell their products properly. For many decades, coal, steel and the products manufactured from them were distributed rather than sold. Special customer requirements did not play a large part, because we were in a seller's rather than a buyer's market. Now the transition is coming hard. These companies, too, can only be reached with difficulty.

There is another damaging influence at work, one which is not peculiar to old industrial regions, but to the German economy as a whole. Here, companies are run purely on the basis of management economic principles. There are scarcely any managers who appreciate that they are in the middle of a social environment and must commit themselves to this society. They are neither able nor willing to think politically or to act in a politically responsible way. In Germany this is the responsibility of others. In my experience, it is therefore rare to find an approach which considers more than purely business events. It is difficult to find companies that are willing to participate in discussions on the future economic development of our Land.

Now on to the organisational criteria. In Germany we have a social market economy. This means that the state can stop investment, but it cannot force anyone to invest – and that is how it should be. However, investment is crucial to a successful structural change. The state has therefore begun by creating financial incentives. It promotes investment with not inconsiderable funding – currently up to 18 per cent of the total investment. But, in the final analysis, this financial aid does not control investment. Most companies will only take it if all other locational factors are right. Since 1979 the Land government has also been trying to transfer know-how to companies. But with all its successes, this strategy has still failed to live up to expectations, for the reasons already mentioned.

The future

What remains is the search for new forms of dialogue and co-operation between the state, institutions and companies. We do not want a system which commands and controls from the centre – we want conviction and co-operation. We have based our structural and industrial policy in the regions of North-Rhine Westphalia on these essential principles.

I am optimistic that companies are becoming more willing to co-operate. What else should I expect in a market economy? I cannot think of any other solutions in such a system. But this process requires infinite patience and time. We have to think in terms of decades. The past 25 years do not seem to have been enough, in any case. But the unemployed do not have this time. Therefore, structural policy cannot be the only solution to our economic problems. What is needed is a net of social provision. But that is another and wider subject.

10
'PRODUCT DEVELOPMENT' AS THE NEW FUNCTION OF BUSINESS PROMOTION

DR RAINER BUHR

Business promotion: the traditional approach

Anyone who is the least bit familiar with business promotion will know that there is a whole series of measures for promoting enterprise or entrepreneurial initiative. This begins, on the one hand, with the rather controversial claim that nothing should be done – companies should be left to their own devices and, if possible, the only provision made should be that no obstacles will be placed in the way of the business activities concerned. It continues, on the other hand, with offers of active help, such as advice on setting-up, financial assistance, product and marketing guidance, provision of favourable facilities in industrial parks, promotional and technology centres and also the allocation of industrial areas, etc.

These incentives are, among other things, also an element of what the Land government offers as part of its package of measures aimed at coping with the processes of change involved in structural transformation.

We, however, as the Land's economic development company, make less use of these measures. This has to do with the fact that we are seldom asked for help purely with the foundation of businesses and there has recently also been a significant drop in the number of those who want to find out from us where they can find suitable production or service premises. Basically, in so far as the traditional, formal measures of business assistance are concerned, we are only used to the extent that we are expected to provide information on possible promotional programmes run by the Land.

Opening up export markets

What we do, however, and this is more an indirect contribution to providing incentives for enterprise, is to develop additional markets, specifically foreign markets, with our current focus being Eastern Asia. We work on the assumption that measures designed to prepare the market and the resulting indications of concrete demand from abroad will provide incentives for increased entrepreneurial activity. However, this does not relate so much to the formation of new businesses, as to the quantitative, and sometimes also qualitative, consolidation of existing entrepreneurial activities. In short, we assume – probably with some justification – that export opportunities create good conditions for entrepreneurial activity.

As a Land company we do not tend to concentrate our foreign activities on giving individual companies export aid for particular products. This is because, on the one hand, we are obliged to display a certain neutrality and, on the other, there are enough specialist institutions operating in this field. Instead, our function is to bring North-Rhine Westphalia as a whole closer to the foreign target region concerned and its institutions, in order to create the marginal conditions there, on site, under which individual entrepreneurial activities can develop. Of course, at the same time, we keep one eye open for possibilities of attracting investment to North-Rhine Westphalia.

However, we have been noticing for some time, both with our market development schemes and also when looking for business abroad, that the traditional subjects, presented in a great variety of ways, are having less and less effect – if they ever did have any effect. What I mean by 'traditional subjects' is the spreading of information on our Land in general, on its economic, social and political structure, on its companies; in short, on its advantages. To put it another way, for some time now we have been looking upon the Land as a product to be marketed; it then emerges that, although we have always encountered a certain amount of interest, it has generally been among the wrong people. To put it in rather controversial terms, lectures on the subject 'Why is it so lovely on the Rhine?' interest the sort of people who always attend such lectures, but whose functions, as a rule, are limited to the 'wining-and-dining circuit' and who very seldom participate in the actual decision-making processes of companies or public institutions. We concluded from this that in so far as our foreign activities – and therefore also the indirect promotion of our business activities – were concerned, we needed another marketing product, namely, one which would arouse the interest of the right people – the decision-makers or, still better in some cases, the decision-planners.

Enabling business closures

However, the need for a new product and, therefore, the need to develop one has also emerged in connection with another aspect of our activities. Of course, we not only operate abroad, but also in North-Rhine Westphalia itself. It has become increasingly evident of late that business promotion not only means helping with the formation or location of businesses or else monitoring the number of businesses, but that business promotion and economic development can, and unfortunately must, also mean helping with the closure of businesses. We have recently been used more and more to co-ordinate measures culminating in the removal of businesses from the market, in a way which is socially and politically acceptable.

With this sort of business promotional activity – we call the associated programmes 'retreating projects' – there are always two major subject areas which have to be dealt with. The first relates to direct assistance for those employees affected by the company closures in each case. This involves questions of welfare planning, of provisions made for the individual by the employment authorities. However, it also involves the training of the employees concerned, and this brings us straight on to the second subject area, namely, the question of which new company activities we have brought in to replace the old ones.

This in turn has brought us to the problem itself. At the end of the day, the fact that a company closes is indicative of a failure of entrepreneurial performance, whatever the reasons for this failure may be. On the whole, it is objective market conditions which force the companies concerned to heavily restrict their activities or suspend them entirely. However, these market conditions are often associated with errors on the part of management, which in many cases was unable to assess the consequences of foreseeable developments.

Consequently, incentives for the development of new business initiatives can only be expected from this side under very limited circumstances. Instead, suggestions on what action should be taken now or which economic activities should be pursued are expected from the public bodies, either from local government or, in many cases, from the Land government too. In most cases this expectation has been fulfilled, too, mainly through measures being taken to build up or improve the company-related infrastructure. In this way, many transfer points, promotional and technology centres and industrial parks and similar establishments have been created. This should all help to create incentives for entrepreneurial activity by establishing favourable conditions. The use of these measures has also produced the desired positive effects. The framework produced by the public sector infrastructure schemes has stimulated new business activities and a wealth of new jobs have also been created.

But even these measures are now being stretched to the limits of their effectiveness. To put it rather simplistically, we cannot build a technology centre wherever a factory closes down. We have to come up with something else. But what?

Product development

We have now arrived at the focal point of the subject of this chapter: 'Product development as the new function of business promotion'. In view of the processes of structural change which we are currently experiencing, business promotion is being increasingly confronted with the need not only to contemplate the marginal conditions of business activity, but also its content. In other words, at the current stage of structural change, the action fields of business activity are not necessarily opening up to the businessman automatically; instead, they are having to be opened up and developed by third parties – often those involved in business promotion. To put it simply, the company has to be told what to do and what the result should be. This is what is called product development in business circles.

So what is 'product development'? Product development certainly does not mean helping individual firms to design products, neither does it mean providing help with individual company plans. Although the latter is one of the activities of business development establishments, it is primarily dealt with by local government business development agencies, and is not a matter for a Land company.

According to our definition, product development means:

- identifying needs;
- checking the potential marketability of these needs;
- producing formulations to satisfy these needs;

- condensing these formulations into 'product packages'; and
- initiating and pressing ahead with processes designed to achieve these product packages.

Anyone involved in this field will know that the main difficulty lies in defining needs, both in terms of content and scope. I must admit that we have not yet made much progress with systematic approaches in this area. However, we are in the process of designing, in close collaboration with the Ministry for the Economy and other establishments, methods of systematically registering the future needs of our Land and, naturally, other areas of the world. Nevertheless, we have got on rather better in practice than in theory. I will therefore use an example to illustrate how the product development process runs. This will also clarify the specific content of the rather nebulous theoretical terms.

A case study: recycling obsolete motor vehicles

In 1991 the Federal Environment Minister announced a decree on the repurchase of old cars. This repurchase decree reflects a general trend in environmental policy, not only to hold producers responsible for the manufacture and distribution of their respective products, but also to extend this responsibility to the proper removal of products when their service life has finally come to an end. Proper removal in this context means that it is no longer enough simply to dump or burn the rubbish or scrap. The consequence of this is that recycling should be integral to the disposal of obsolete products.

The problem which now emerged in the case of the car industry was that, on the one hand, the technical, logistical and financial conditions required to build up a functional recycling industry were not present but, on the other hand, a solution to this problem had to be found within a relatively short period. The car industry in North-Rhine Westphalia therefore approached the Land government to ask for help in finding a solution to the problem under consideration.

As the Land's business development company we, along with the Ministries for the Economy and the Land's Ministry of the Environment, and also the Ruhr Initiative – an association of companies from various branches of industry aimed at promoting economic development in the Ruhr – assumed the task of organising a working group. Its job was initially to create a development plan for an integrated recycling system and then, as the next stage, to implement this plan step by step. The special thing about this working group was that it included companies from the car, chemical (plastics and rubber), disposal, glass, metal-shredder and dismantling industries. It was thereby guaranteed that the development plan would not be an abstract notion, but would be based on the knowledge and experience of the companies concerned. There was therefore no doubt that the desirable and feasible could be largely reconciled. The development plan was presented last autumn; the first steps towards its implementation are being taken now.

By collaborating on this project, we ourselves have learnt that there is a need here which probably goes far beyond the question of removing old cars. There are signs that a whole new industry, with a relatively simple *raison d'etre*, will grow out of this need.

To date, ever more complex systems have been developed for manufacturing products. Old products were removed, and to some extent still are, based on a principle which even our forefathers in the Stone Age were familiar with, namely, throwing away or burning. In future, the same intelligence, the same technology, the same logistics that have been invested in the development of products will have to be invested in their disposal too. At the same time, this will affect the manufacture of future products.

We have summed up this emerging development under the term 'retroproduction'. We are clear, in this respect, that the elements which constitute this term – the environmental problem of waste, waste industry, recycling, new production processes – are not new. The only new thing is the way in which these elements, and their orientation towards the goal of developing a new industry, have been brought together as an abstract concept. As with production, retroproduction too requires co-operation to be organised between different branches: there is a need for new processes and new services. This once again presupposes an increased need for research and development. Each of these aspects can give rise to partial measures and therefore to new business activities.

In the meantime we have started to use retroproduction with its car recycling subcomponent in measures aimed at safeguarding sites during company closures. We have noticed that the subject has been taken up and pursued with great interest, not only by politicians, but particularly companies and businessmen too (in the meantime, the second retroproduction project, namely, electronic scrap recycling has also been launched with the participation of the branches concerned).

So what have we done with regard to 'product development'? We started out with a problem (waste as an environmental problem) which incorporated a specific requirement (recycling measures). Based on the results of development studies on car recycling, we became convinced that, viewed long term, this was a need with marketing potential. We summarised the formulations designed to satisfy this need under the concept of retroproduction or the recycling industry. This concept is our product. And this product will give rise to concrete measures which, in turn, will provide further incentives for entrepreneurial activities.

However, to prevent these incentives from remaining purely theoretical, it is once again necessary for the business promotion measures which have been brought along, as I mentioned at the start, to be put to use. This means financial support, it means advice and it means training. The Land government is currently in the process of adjusting its promotional instruments, so that the 'product' presented here finds its 'consumers' too.

Finally, let me briefly return to our foreign activities. Here, too, it has emerged that products of the type described earlier have met with considerably more interest than traditional site presentations. So, at the end of March this year, we took part in a specialist exhibition for car recycling in Tokyo as the Land company, where we presented our retroproduction 'product', Within two days we had received over 3,000 visitors at our stand and the staff in our Japan Department are still busy today with the follow-up to this campaign. We are extremely pleased with this result, as it shows that sensible product development through business promotion not only benefits our clients, but also gives us a strong incentive to continue doing something.

11

THE WORK OF THE TRAINING AND ENTERPRISE COUNCILS

PAT RICHARDS

TECs and pit closures

Training and enterprise councils or TECs were launched three years ago. It was the Government's ambitious bid to involve business leaders in running its training and enterprise, and innovation and education programmes. There are some 82 TECs up and down the country. We handle something like £2.3 billion worth of public money every year. That money is handled by independent companies run by business men and women.

We are focused not just on mining and on individual ex-miners, but on a wide spectrum of sectors and individuals. The principal aim of the TECs movement is very simply this: 'To improve business performance, primarily through the more effective use of people, to contribute to local economic development'. Because we have that particular remit, in October 1993 the Secretary of State for Employment was given, in addition to the £2.3 billion that we handle every year, an extra £75 million as special assistance for coalfield communities. She asked the 10 TECs in areas where closures were announced to put together strategies, action plans for handling those closures. We were given two weeks to put those plans together. Most TECs responded very well to that invitation, although what I will do is focus on the work we did in North Nottinghamshire.

The context of the local economy

North Nottinghamshire is an area of some 50 square miles; it has, at the moment, 13 pit heads in that small geographical area, eight of which are threatened by the present closure programme. Three of those have already closed.

We spent two weeks, 24 hours a day, speaking to as many people as we could, holding workshops, meetings, dialogues. Employers we saw as a particularly important partner, and we wanted to discuss with them what they thought their priorities for action were. We wanted to ensure that they were involved and we had their commitment from the outset. There was a tremendous commitment which we have certainly been able to maximise. We worked closely with local authorities and with English Estates, which is the property arm of the Department of Trade and Industry, in order to discuss how our plan would complement and reinforce their own proposals. What we did not want was just to bring in our £7 million share of that £75 million and start to reinvent the wheel, but to make sure that it complemented and reinforced what was already happening. And of course we worked closely with British Coal Enterprise and with the public employment service. We also spoke to individual miners; we wanted to gauge their reaction, we wanted a customer reaction to what we were proposing and we wanted their views. And, in addition, the trade unions – the National Union of Mineworkers and the Union of Democratic Mineworkers. We never managed to get them round the same table but we did talk to them separately.

We already had particular problems in North Nottinghamshire. These can be summarised as:

- an overdependency on coal-mining and other traditional industries;
- a fragmented and under-developed service sector;
- a very poor track record in attracting inward investment;
- high and persistent levels of (increasingly long-term) unemployment;
- poor ground conditions and environment; and
- poor East–West communications and peripherality from expanding (and especially European) markets.

We had, for some time, an overdependency on coal-mining and traditional industries, including textiles and industrial engineering. We did not have a very well developed service sector. Our companies, whether they be manufacturing or service sector, tend to serve local markets. Forty per cent of our companies are serving a North Nottinghamshire market – not a very healthy position to be in. We did not have, as an area, a very good track record in attracting inward investment; we never had to, because we already had sufficient jobs with British Coal.

In 1989 something like 10 per cent of all jobs were in the mining industry, which amounted to some 14,500–15,000 jobs. That has now dropped to something like 5,000, just three or four years later. We do have high levels of unemployment. The average is about 11 per cent but there are areas where it touches 19 per cent to 20 per cent: pockets of degradation. Environmental issues are there to be seen, because of past mine workings and inadequate investment in land reclamation. The infrastructure in terms of East–West communication is very much in need of improvement.

The impact of pit closure on the local economy

This is the local economic context in which pit closures occur. The specific impacts of closure are, in the short term:

- 3,000 jobs lost October 1992 – June 1993; and
- £48,000,000 lost income.

and in the longer term:

- loss of orders in engineering transport and power generation;
- reduced expenditure on local goods and services;
- further voluntary redundancies;
- continued rise in the jobless total; and
- increased business failures.

We have lost 3,000 jobs since October and we have lost £48 million in income. If the average salary of a mineworker was of the order, including bonuses, of £16,000 per annum, that is £48, 000,000 of lost income no longer circulating in the North Nottinghamshire economy. The average age of the redundant worker is 35. These are not old men, they are in their mid-thirties: They have young families, they have mortgages, they have particular other problems that need to be addressed. In the longer term, one to three years certainly, that multiplying effect, that ripple effect, will be felt in terms of lost orders and engineering transport and power generation. We are already experiencing reduced expenditure on local goods and servicing. Voluntary redundancies among suppliers and sub-contractors to British Coal are also very much in evidence.

We are also aware of what might happen locally unless there is effective local intervention: there will be a continued erosion of the employment base; a loss of individual, community and business confidence; a diminishing number and range of investment opportunities. These economic factors would have social consequences: falling house prices and mortgage indebtedness, greater out-migration and even an increased level of crime.

The strategy of North Nottinghamshire TEC

So we had a sense of what closures would mean, economically and socially, in the long and short term and the catastrophic consequences we wished to avert. So we looked at strategy, which has three parts. The first I would like to focus on in more detail, but I would also like to cover the other two.

Objective one is focused on regenerating and modernising, and helping local communities diversify their product base. What we haven't aimed to do is to engage in a massive business start-up programme. We've recognised that the greatest need is at that small- to medium-sector end, and that it is existing companies who need help to survive and to grow. Although we do have a small firms start-up programme, we are not financing it at the expense of medium-sized companies. What we want very much is an economy which is robust, an economy which is diverse. We do not want to get into the situation again where we have a dominance of one industry and the vulnerability that that exposes us to. The Secretary of State invited us to put together an action plan and, I have to say, I think they were thinking more in terms of training, training, training and training. Their idea was that all you need to do to help unemployed miners get back to work is to give them a six-week conversion course to transfer their skills to another sector, or some up-skilling course. We have said categorically 'no'. The emphasis has to be on helping to create and protect jobs. Labour market activities are more effective than direct help to unemployed people. And that's a message that we keep hammering home, though it is not heard by everybody.

The second objective is about community regeneration: making sure that employment support, and training support, also addresses social issues, making sure that we involve residents in the renewal process in order to protect and enhance the value of the investments made but also in order to make sure that we maximise the use of all resources. People do tend to think of resources in terms of money. There is a lot of commitment among the people in North Nottinghamshire to help themselves and we need to maximise that commitment. We want to ensure that the agencies that are already in existence, the voluntary agencies, the statutory agencies, work well together. So we have got programmes to co-ordinate and stimulate networking.

The third of the objectives focuses on individuals, and is about helping them to overcome the barriers to employment. We recognise that miners do not all have readily transferable skills; they tend to have limited job-search attitudes, and they tend to look for employment in a narrow range of secondary occupations, like factory work and warehouse work, precisely the sorts of jobs which are vulnerable to future redundancies. We are trying to avoid that limited job-search activity. There are linked problems in selling houses. As I said, most of the redundant miners are in their mid-thirties, they are house owners, they have mortgages and they are not able to just up and go because they have got problems in selling and problems with mortgage indebtedness, even if jobs were available elsewhere.

We have tried to make sure that the additional money available is effectively targeted: that the resources are allocated where they are needed most. The additional assistance is available to all former British Coal employees, their partners and dependants. Other contributors have emphasised the pressures on family life. The fact is that most of the jobs that are available are the types of jobs that women are more likely to accept and men are not likely to want. It was very important to us that the special assistance was not only available to redundant miners but also to their families, to their wives and dependent children. That help is also available to subcontractors and working on-site to British Coal. In addition, every single unemployed person within a ten-mile radius of a pit closure is also eligible for special assistance and every single company across North Nottinghamshire is eligible for special assistance.

The type of assistance available is based on three pits at the moment and amounts to £1,800,000. That sum of money will only be available for 12 months. It is there to focus on small to medium-sized companies, to help them to survive, to diversify their product base. Our strategy includes 'TEC line', employer outreach, enhancing business skills and attracting new jobs. 'TEC line' is very simply a telephone help line, a focal point of contact for business support on a range of issues: health and safety, quality, managing change, exports, etc. We are handling about 2,000 calls a year: we expect that to rise to about 5,500 in the coming 12 months.

'Employer outreach' is very simply a programme which recognises that you can't expect companies to come to you, you have to go to them. They need help on their own premises. We have a team of people who are engaged in business counselling, helping companies to carry out spot analysis, to identify consultancies which will help them to undertake one project or another, to improve their business competitiveness.

Certainly more money is being channelled into research and information: understanding the market-place better; and raising awareness among the business community of the professional services which are available, including a range of financial packages.

What we are looking to do is to work in partnership with the other agencies to make sure that the money goes as far as possible, and to ensure that the funders act intelligently, that they throw away the rule book and look at the sense of what is being proposed.

'Enhancing business skills' includes a programme of courses, open-learning, flexible learning and management training programmes of one kind or another, geared at that small firm sector, focusing particularly on marketing, market research, export skills and financial management.

The last item is about attracting new jobs into the area, looking again very closely – in collaboration with the local authorities, with the Department of Trade and Industry and with other agencies who also have that particular focus to make sure that North Nottinghamshire is seen as an area to be targeted and a targeted marketing campaign is what we are putting together at the moment.

PART V

ENVIRONMENTAL ISSUES: LAND RECLAMATION AND LOCAL REGENERATION

12

WASTE PRODUCT REHABILITATION IN MINING UNDER THE CONDITIONS OF THE EMPLOYMENT CRISIS IN THE FORMER GDR

ANDREAS WIEDER

Introduction

Mining everywhere is associated to widely varying degrees with the use of the landscape. This quite clearly applies to open-cast operations. Large holes are more obviously foreign to a landscape than buildings or harmful substances concealed in the ground. The risks associated with these waste products can vary enormously. But even when the legacies of former mining activity are relatively harmless, the following still applies: 'Before the site can be used again it must be reconditioned.' Here, different requirements apply to underground mining in Germany – primarily involving hard coal, iron ore, copper and uranium – than to the old, predominantly brown coal, open-cast mines.

The aim of this short chapter is, firstly, to show how the immense task of rehabilitating the waste products of former mining sites is being tackled in the former GDR and what has been achieved there to date. The second aim is to show that this is possible, even when the state is in a difficult financial situation and companies are faced with a severe structural crisis, provided there is a readiness for conflict, but above all, a great willingness to co-operate in order to achieve a detailed end result.

The legacy of mining in the GDR

The rehabilitation of former mining sites is made more difficult by the fact that the mining industry itself was completely oversized in the former GDR. This was partly due to an attempt to be as self-sufficient as possible with regard to raw material and energy supplies from the world market (brown coal) despite poor deposits; and partly due to special supply requirements from the former Soviet Union (uranium) despite the costly technical difficulties involved in its extraction. With the opening of the borders and the end of the GDR which followed shortly afterwards, the practice of operating mines based on the objective of self-sufficiency in energy ceased.

Coming at the same time as the severe structural crisis in the new Federal Lands, this led to a particularly far-reaching crisis in the different mining sectors. This resulted not only in a loss of employment and the closure of individual pits, but in many branches of mining and mining regions being completely abandoned. This led to a situation in which a large amount of former mining land became available within a very short time and this had to be rehabilitated. At the same time, there was a huge legacy of former mining sites from recent years and decades in the GDR, which had not been adequately reconditioned due to the cost factor.

Then there were mining companies which were still in a very difficult situation, due to the structural crisis and contraction, and were unable to organise and finance this process of rehabilitating their waste products independently. The crisis among mining companies was accompanied by a huge loss of jobs. Whereas the number employed by the largest mining sector in the new Federal Lands, that of brown coal, still stood at around 133,000 in 1989, by the end of April 1993 it had shrunk to 51,800, as a result of closures, contraction and rationalisation measures.

Copper and iron ore mining, which was only of regional significance, was completely abandoned, as was uranium ore mining. The uranium ore mining company, Wismut GmbH, now operates almost exclusively as a redevelopment company, following the elimination of some parts of the company. Potash mining has lost roughly half its production. Of an original workforce of roughly 30,000, only about 6,000 still have their jobs today, following massive simultaneous rationalisation.

This crisis deprived companies of the opportunity to fulfil their obligation to recultivate areas of waste products from their former mining sectors and, at the same time, reduced the potential for finding a subsequent use for former mining sites in industry, agriculture, house building or as recreational areas. These two factors together threatened mineworkers with the collapse of their employment market. Employment opportunities at their former workplace had disappeared or had at least been massively reduced and new businesses were not relocating there due to a shortage of suitable land or the many abandoned buildings.

Policy measures

New ways had to be found of rehabilitating closed mining areas in the new Federal Lands, which would have bankrupted the surviving companies in no time, but also, at the same time, of improving job opportunities in areas seriously affected by structural change. It was initially decided that all land remaining from closures made before 1st July 1990 should be treated as constituting waste products, for which the companies were no longer financially responsible. A subsidy was made available through a Federal government programme to prevent the companies' reserves being used up through reclamation costs. At the same time, finance was guaranteed for the recultivation of land and disposal of waste over a transitional period of a few years. The financial requirement for brown coal alone is about one-and-a-half billion DM a year for five years: 75 per cent of this will be covered by the Federal government and 25 per cent by the Federal Lands concerned. Apart from this, relatively little funding comes from outside finance. The companies themselves mainly supply machinery and tools for the work.

Rehabilitation work was given a very high priority to begin with, as closures, and with them job losses, increased as the structural crisis spread. It was, and still is, almost exclusively carried out by former mineworkers. To begin with, this took the form of job creation schemes. This is an instrument for the occupational reintegration of the long-term unemployed which has been around for years. Anyone who has been unemployed for over 12 months can find a new job via this route. The employer engages the new employee in activities which would not otherwise take place (which means that this is not replacing a normal job) and receives up to 90 per cent of the wage paid from the employment office. Such job creation schemes are subject to certain statutory modifications in the former GDR. The qualification period of unemployment does not apply and the wage subsidy paid to employers by the employment offices can be up to 100 per cent, although this is based on a working week which is up to 90 per cent shorter and is therefore a lower wage overall. The job creation schemes are limited to one year, or two years in exceptional cases. However, almost all schemes have lasted two years in the former GDR. For two years in 1991 and 1992, these job creation schemes protected up to 400,000 people in total, not from redundancy, but from the experience of long-term unemployment, following the economic collapse in the former GDR. At the same time, the many thousands who worked on these schemes in mining have completed, or at least pushed well ahead with, several rehabilitation schemes during this time. They have thus opened up many opportunities for environmental improvements and new economic and social developments in areas with a hitherto lop-sided structure. With the conclusion of many job creation schemes, the remaining mining companies have founded redevelopment companies, which can now operate for five years using the financing already described. Sixteen thousand employees will be engaged in rehabilitating brown coal waste products, but for many of them it will not be for the entire duration of the company.

The Mining and Energy Union had to enter into separate collective wage agreements for these employees, which were specially designed to meet the needs of this particular group. Incomes in brown coal rehabilitation are roughly 70 per cent of those employed in the active brown coal industry; if the so-called instalment payments are included, they are roughly 90 per cent of the collective wage agreed for comparable employees. It is unfortunately the case that the full wage comparable with mining employees cannot be achieved for legal reasons.

The impact of policy measures

A large amount of work has already been done on the reconditioning of mining waste products. In one redevelopment area alone, the central German brown coal district around Leipzig, over 70 million cubic metres of coal waste were moved by approximately five-and-a-half thousand employees and 1,167 hectacres of land thereby recultivated. This involved, for instance, dismantling almost 260 kilometres of railway track and over 36 kilometres of conveyor belts. Lakes and local recreation areas are emerging, as well as conservation areas and new industrial estates. Comparable developments are underway in other redevelopment areas too.

The financing of these redevelopment companies is, as already mentioned, guaranteed for the next five years for the largest sector, this being the rehabilitation of brown coal waste products. This timescale has been set, not so much in the hope that all redevelopment work will have been completed in five years, but primarily due to the uncertainty surrounding brown coal privatisation. A fresh decision will then have to be reached concerning any waste product rehabilitation work still outstanding. In contrast to this, recultivation in the area of active and recently closed mining operations is the legal responsibility of mining companies.

A rapid rehabilitation of mining waste products was urgently required for various reasons. Firstly, on account of the many industrial ruins covering the landscape, which made it less appealing to new investors. In addition to this, they were preventing numerous pieces of land from being put to a new use. The final reason was that the economy in the former GDR had shrunk massively and a huge employment crisis was developing.

This development was made possible, after the government had waited for several months, by the fact that the trade unions had lobbied effectively. However, the question of whether this government's policy in the former GDR was adequate was also becoming increasingly contentious within the government parties. It was not so much the trade union campaign which made such a massive redevelopment possible, but rather a change in government policy as a result of outside pressure. After this phase, all sides concerned displayed a great willingness to co-operate, in order to achieve compromises acceptable to everyone on the details of the policy.

Neither the financial provision nor the workers' terms of employment are completely satisfactory to any of those involved. The government wanted more private and less public financing; the unions wanted better employment and living conditions for the employees there. But, all in all, it is a result which has for some time been tackling two important problems – the employment disaster in the economy of the former GDR and the removal of its legacy of industrial ruins – and with some success in both cases. In the present economic crisis, it also gives employment prospects to many people in the former GDR which, due to the size of the task, will probably extend beyond the five years described. Therefore, this policy is viewed by all sides, or, in any event, it has been to date, as a clear victory for common sense.

13

LAND RECLAMATION AND ECONOMIC RESTRUCTURING IN NORTH-RHINE WESTPHALIA

FRIEDRICH WILHELM WAGNER

Introduction

There is a tradition of regenerating and mobilising mining land in the coal-mining areas of the Federal Republic of Germany, due to the prolonged adjustment process in hard coal mining and the many years' experience of extensive open-cast mining in the brown coal mining areas.

North-Rhine Westphalia has always been faced with a particular challenge in this field, since its border areas are home to both the Ruhr, Europe's largest hard coal mining area, and the Rhine's brown coal mining district, taking in an enormous area of land. The tasks of regenerating former mining land have proved completely different in the two sectors, simply due to the general differences in mining technology used in underground hard coal mining in the Ruhr compared with the Rhine's brown coal mining, which has been characterised by large-scale, open-cast mining since the Second World War.

Hard coal mining in the Ruhr

Let me begin with a brief summary of the Ruhr and hard coal mining. Since Germany's first hard coal crisis came at the end of the 1950s, North-Rhine Westphalia has experienced the complete restructuring of mining regions. If we consider the period from 1957 up to the second oil crisis in 1979/80, 85 working hard coal pits with a saleable output of around 64 million tonnes were closed down in the Ruhr. The entire small-scale mining industry on the southern edge of the Ruhr, comprising around 164 small-scale working collieries, disappeared during the same period. These figures reflect the enormous blows to the Ruhr's mining industry in only 20 years. The concentration of the mining industry and cuts in capacity have represented an immense challenge to employment market, social and structural policy. Since the start of the first coal crisis in 1957, almost 400,000 mining-related jobs have been lost in the Ruhr. As a result, the primary sectors of coal and steel have lost their position of priority in the industrial landscape of North-Rhine Westphalia.

This transformation of the Land from a region characterised by coal and steel to an export-oriented, hi-tech centre with coal and steel has been achieved not least thanks to the early assistance and backing received from North-Rhine Westphalia, particularly with regard to the mobilisation of former mining land.

Intervention by regional government 1966–1988

Back in 1966, the action group, Deutsche Steinkohlreviere GmbH, was created along with industry and with the help of the Federal government. The purpose and aims of the action group were:

- to grant compensation for the closure of hard coal pits;
- to use property belonging to the closed mining companies for the location of businesses and the implementation of restructuring measures; and
- to grant mining damage assistance to companies wishing to set up new businesses on property damaged by mining.

The action group's property marketing operation had acquired and used around 17,000 hectares of disused land before its liquidation in 1988. Just 4,000 hectares of this were used for new industrial or commercial purposes. Land uses have varied from the creation of countryside parks, through the location of shopping centres, businesses and house construction, to an industrial museum which is classified as a historical monument. As a rule, it took 10 years from the planning to the completion of the schemes. The crucial time factors in this respect proved to be: the time-consuming removal of extensive spoil banks and building remains; infrastructural developments and connection by means of roads; supply and disposal facilities; and the inspection and rehabilitation of subsoil contaminated by waste products.

The current hard coal situation

Under the latest coal policy rulings issued by the Federal Republic of Germany on 11 November 1991, North-Rhine Westphalia is now once again facing a huge challenge involving the structural transformation of its coal areas. The German hard coal mining industry will have to reduce its sales to power generation and the steel industry to 50 million tonnes per annum by the year 2005. When the sales potential available on the heating market is taken into account, this produces a total capacity for Germany's hard coal mines of roughly 55 million tonnes per annum in the year 2005. This can be compared with the production figure for 1990 which was roughly 70 million tonnes of hard coal. Implementation of the coal policy resolutions will result in the loss of some 30,000 jobs through the closure and merging of pits. Mining companies believe that by 2005 five pits will have been closed, four will have had to reduce their output considerably and three more will have merged with neighbouring collieries to produce integrated operations. In other words, of the 19 pits currently working in the Ruhr, 12 will remain.

Based on its experience of the last 30 years, the North-Rhine Westphalian government will in future continue to adopt its dual strategy in this context. This will involve, firstly, giving hard coal a long-term future perspective based on reduced output and, secondly, forcing structural change in those areas which will see mining retreating still further in the short and medium term.

The aims of structural change are divided into three main elements:

- to modernise the region's traditional branches of industry, coal and steel, and save jobs without impeding structural change;
- to create jobs in new, up-and-coming branches of industry (e.g. environmental protection, high-tech); and
- to make this process socially compatible and regionally balanced.

Land reclamation in the future

Above all, what will be crucial to the success of the structural change in the coal areas will be the transformation of land left derelict by closed collieries for the location of businesses, for residential areas and leisure and cultural facilities with the appropriate transport infrastructure.

It can therefore be expected, based on the merger and closure plans of the mining companies, that by 1997 the amount of former mining land available in the coal mining region will have increased by around 1,300 hectares. Since there is otherwise a shortage of space for industrial location in these regions, the restructuring of the coal areas will only be possible if this huge existing potential of disused industrial land is reclaimed. Sixteen former sites, covering a total area of around 500 hectares, have already been converted to other uses or have been the subject of planning applications.

When reclaiming such land, the most important factor is the link between intended future use, the legal planning requirements involved, and its appropriateness to the quality of the site, especially the disposal of waste. In principle, the problem of waste products must be investigated in each individual case before such land is marketed, in view of the different nature and standard required of each site. As a rule, investors are only prepared to talk about specific projects if the necessary planning permission has been obtained and the question of waste products resolved. Due to the possible risk associated with waste products, former mining land is considered to be less valuable by potential users. Additionally, there is the problem that users should be protected from any future obligations for land reclamation. So, apart from the aim of bringing an ordered structural change to this disused land, planning must conform to the rehabilitation expenditure needed in each individual case and the environmental risks which may remain.

What must be particularly emphasised is that, unlike all other areas of disused industrial land, when land which has been used for mining is restored, the mine-owner has a statutory obligation to regenerate it under the Federal Mining Law. This involves, firstly, evidence of the elimination of any possible contamination of the land. The removal of such acute hazards is the purpose of the final operating plan, which must be drawn up and implemented by the pit owner. The operating plan is approved by the mining authorities, which also oversee its implementation. A more extensive waste product clean-up operation, as required for regeneration, must conform to whichever records exist for the area concerned which are binding under planning law.

The instructions contained in the final operating plan may relate to such measures as: the treatment of ground-water contamination, e.g. by means of purification plants in the ground-water down current; the excavation of earth from individual areas where there is severe contamination; and, if necessary, thermal soil treatment, e.g. in revolving tubular kilns. Therefore, the need for long-term ground-water treatment on this sort of disused land may mean that it can only be used for gardens or parking areas or for the location of business premises which do not need deep foundations or basements.

The regional framework

It is primarily the towns and communities which are called upon to create a successful structural and development policy. It is they who can themselves determine the extent of rehabilitation necessary for the subsequent use, and therefore the time required, using planning which is adapted to the quality of the site concerned. The ever-renewed call for mines to be obliged to return their land to its pre-industrial state, thereby making it suitable for any type of subsequent use, is not particularly conducive to the rapid mobilisation of former sites, nor is it legally enforceable.

North-Rhine Westphalia has decided, in the interests of achieving rapid structural change in the coal districts, to promote the formation of regional development companies, whose job it will be to plan and execute individual development projects within the region. The partners in such development agencies may be the landowners – usually the mining companies – the towns and communities or other interested investors. An example of this is the recently founded Eastern Ruhr Regional Development Agency.

North-Rhine Westphalia has promised its towns and communities the opportunity to secure aid from the Land's budget to cover any planning and development costs for the intended subsequent use which exceed the mining company's statutory obligation. Therefore, at the end of 1991 it was decided that a working framework should be created for coal-mining areas at the Ministry for the Economy, Small Businesses and Technology. This offers resistance to accelerated job cuts in the shape of an even more intensively co-ordinated structural policy, which also incorporates EC community programmes.

However, this working framework, with its substantial funding, goes far beyond being simply a financing instrument. The aim, subject to there being regional collaboration with the coal-mining towns and communities concerned, is to develop and co-ordinate the key areas of: eligibility, mobilisation of industrial land, industrial promotion, traffic infrastructure, the environment, urban construction and dwellings. Several major projects are currently at the planning stage in which North-Rhine Westphalia intends to invest several hundred million deutschmarks. An outstanding example of this is the staging of the 1997 Federal Garden Show on the site of the Northern Star pit in Gelsenkirchen.

The brown coal areas

The problems of regenerating and rehabilitating the retreating hard coal mining industry and its legacy of disused industrial land, some of which has been used for mining for a hundred years or more, are fundamentally different from those in the regeneration and recultivation underway in the Rhine brown coal district. The enormous amount of land required for open-cast brown coal mining in predominantly agricultural areas means that rational regeneration can only be successful if the recultivation has already been established and agreed in detail before the mining intervention begins.

Therefore, North-Rhine Westphalia subjected brown coal planning to a strict statutory regulation early on, based on the Land Planning Law, which establishes the instrument of the brown coal planning process as a separate process of the local development plans binding on the remaining Land planning.

This primary regional planning function is brought to life by the Joint Brown Coal Committee, which implements the brown coal planning procedure independently and democratically for each individual open-cast project and presents the Brown Coal Plan, which must have majority approval, to the Land government for final inspection and approval. The individual goals and measures contained in the Brown Coal Plan are then converted into concrete administrative processes, including, among other things, colliery operations planning processes required under mining law.

The regeneration and recultivation of open-cast mining areas which have been fully exploited occurs immediately after the actual mining of the raw material, step by step, and should result in the restoration of the landscape to the condition it was in before mining began. A balance of must be struck between alternative potential uses of the reclaimed land: for example, for forestry and agriculture, water supply and distribution, recreation, biotope and species protection and transport.

According to the regeneration statistics for brown coal mining on the Rhine, a total of 24,500 hectares had been used between the start of brown coal mining and 1 January 1992. A total of 15,700 hectares – more than half, in other words – have now been regenerated and recultivated. The remaining 8,800 hectares or so accommodate the five open-cast mines currently being worked in the Rhine district. These can produce approximately 120 million tonnes of coal a year. Attempts in future to concentrate these still further into only three major open-cast mines will tip this balance still further in favour of the recultivated areas, while maintaining production capacity.

The quality of the new land will depend essentially on the soil material available for recultivation and finely balanced mass management. The vast loessial parcel which is present throughout almost the entire Rhine mining district particularly favours the success of agricultural recultivation. For the 7,100 hectares or so of agricultural land which have been regenerated to date, the prescribed standard has been for the subsoil to be covered with a layer of loess at least two metres thick. Agricultural cultivation starts on the land seven years before it is handed over to local farmers; this is carried out by farms belonging to the mine operator himself who, furthermore, issues a 25 year quality guarantee covering the recultivated land.

The technical formation of brown coal mining in the Rhine district makes purely selective extraction possible, as well as the transportation and tipping of separate, individual types of earth, such as overburden, loess and forest pebbles. Even temporary open-cast planning permission requires high standards of land restoration, whatever its current state. Any soil imbalances which occur are partially offset by transfers between individual open-cast mines up to 20 kilometres apart, using large belt systems or railways.

Apart from the 7,100 hectares of agricultural land which have already been mentioned, roughly 6,750 hectares of forestry land, 770 hectares for dwellings and industrial locations, and approximately 300 hectares of communications, have been recultivated and handed over for subsequent use within the framework of this extensive mass plan. In addition, approximately 800 hectares of varied expanses of water have been created through the remaining mass deficits.

Thus, the fully exploited and almost completely recultivated southern mining district in south-west Cologne is now an extensive landscape of lakes and meadows and a well-known recreational area characterised by a diversity of woods and lakes, large parts of which are conservation areas.

14

ENVIRONMENTAL ISSUES IN SOUTH WALES

EWART PARKINSON

Introduction

Those who know, and therefore love, Wales may know much of what I am about to say in this chapter. I ask for their tolerance. When I was asked to contribute a chapter to this book I seized the opportunity. It was only when I sat down to write that I realised the difficulties of the task.

So much has been achieved in the past 25 years in the regeneration of the Welsh Valleys. Tribute is justly due to the Welsh Office, the Welsh Development Agency and the Local Authorities. They have made a massive contribution both in creating a new infrastructure and in promoting inward investment by the private sector on an international scale. For the first time in 70 years unemployment rates in Wales are below the UK average.

These are commendable achievements. But there is another side to the story. A document produced by Mid Glamorgan County Council last February says:

The Valleys remain one of the most deprived areas of Britain, with male unemployment close to its 1988 level, the highest proportion of economically inactive people in England and Wales, the highest rate of long-term ill-health and above average levels of disability, and the lowest economic output and average household income per head in the UK.

Indeed the region which traditionally had relatively well-paid male employment in the basic industries has now more women at work than men, even if much of that work is part-time and/or ill-paid. The document also said:

Most Valley towns were built over a century ago at the peak of coal mining activities and there has been relatively little investment in their fabric since then. As a result many valley buildings need major refurbishment to bring them to contemporary standards and with new uses . . .

The dilemma is this. How do you promote and market a region with clear economic and environmental potential and yet at the same time campaign for greater resources and greater action to increase the number and, even more importantly, improve the quality, of the jobs in a region where three-quarters of a million people live?

Background

It was here in Wales, and in parts of England, Scotland and Germany, that humanity's historic shift from an agricultural to an industrial society first occurred. That shift is still happening in South America, South East Asia and China. Non-ferrous metal working had developed in the western valleys some 400 years ago and the iron and coal industry gathered momentum in the eighteenth century. The story of coal in South Wales is one of dramatic growth and decline. Regardless of political and economic changes in the UK, the growth in coal production and in the numbers of people employed grew steadily until 1915 when some 260,000 men were employed. The decline, again independently of political and economic changes, was equally relentless to the point at which coal-mining has, for all practical purposes, disappeared from the scene.

A Government committee, looking at the problems of the South Wales Valleys in the early 1920s, proposed the building of a new settlement at Llantrisant to provide better housing for those working in the Valleys' coalfields. In South Wales there are four recognisable sub-divisions: the most northerly and highest division where the coal measures begin to run out, known as the Heads of the Valleys; the Valleys themselves which run on a north–south axis; the places where the valleys open up to the coastal plain, the Valley Mouths; and the coastal plain with its six ports. Llantrisant lies at the Valley Mouth of the Ely river.

Nothing came of this proposal because the decline in jobs (rather than the expected growth in jobs) was beginning. The idea of a New Town at Llantrisant remained, however, in the Government's mind. It re-surfaced strongly in the one attempt by Government to present a detailed set of development policies: 'Wales – The Way Ahead' in 1967. It proposed a New Town of about 250,000 people at Llantrisant to provide new employment opportunities (an intriguing reversal of the 1922 objective) for the Valley people.

A major public inquiry into the proposal was held in the early 1970s. The County Councils supported the Government whilst the district councils of the Heads of the Valleys and Cardiff City – the capital city of Wales – opposed. I was asked to be the expert witness for the City Council. The basic argument was that the policy of growth at one of the valley mouths was a fundamentally wrong policy, especially when all major investment would be focused into one location. Investment resources for infrastructure creation were limited: the proposal would leave the existing network of Valley settlements becoming even more deprived of funding. The better strategy would be to identify a number of towns in South Wales capable of playing the role of growth centres. Investment in these growth centres would be aimed at restructuring the economic, social and physical bases of the towns. This would enable the present scattered settlement pattern characteristics of South Wales to be reinforced. This would give greater recognition to the social solidarity which again is a powerful characteristic of coalfield communities throughout Europe. In the event, the Secretary of State, with great reluctance, accepted the Inspector's advice to reject the proposals. Unfortunately, no clear alternative strategy was adopted. The policy of growth in the Valley Mouths was still the general policy of the Welsh Office, never being officially abandoned. I do not think any clear statement has since been made (certainly not in the 1970s or early 1980s). It was therefore good to hear the Secretary of State say in a speech last January: 'But that approach (i.e. a more comprehensive approach) has to be balanced by giving opportunities to all local communities to take action. I do not want any forgotten valleys.'

In the late 1960s the local authorities combined to campaign for the early construction of the M4 Motorway across South Wales. It was pointed out at the time that the Italians had already built the Autostrada right through the Mezzogiorno to Sicily. But the usual arguments about lack of demand and low cost-benefit were used to delay the project. Now that it is nearing completion (30 years later!) it is being hailed as the 'Corridor of Opportunity'. Certainly a series of major employment opportunities are being created along it.

Another remarkable local authority initiative was the development by South and Mid Glamorgan County Councils of a Valleys Passenger Rail network using the old lines. In the space of eight years – 1984 to 1992 – some 23 new stations were created to make a total network of 73 stations serving a population of about a million. In the early 1980s British Rail was distinctly cool about the concept: now BR Regional Railways is adopting a most positive stance. Some nine million passenger journeys are made each year.

In my view there are two investment decisions that would transform the image of the Valleys and the prospects for a quality regeneration programme. One is to up-grade this rail system to a state-of-the art light transit system with the most sophisticated signalling and information systems of the highest international quality. To those who say that the idea is absurdly expensive, I would give two replies. The first is that I was ridiculed when I originally put the idea of the current rail network arrangements before the County Council nearly 20 years ago; the second is that the public purse found some 200 million pounds to place the A55 Expressway in a tunnel under the Conwy estuary for sound environmental reasons – we are probably talking of the same order of expense for the ten-year strategy to complete an adequate highway system for the Valleys.

In July 1988 the Secretary of State announced a Valley Initiative: a three-year programme bringing together community development, education and training, transport, housing and health. This was extended by his successor, David Hunt, until March this year. He has appointed consultants to assess its impact. That will make interesting reading. Perhaps it will say, in summary, that a great deal was achieved but that a more comprehensive, a more clearly articulated and a much longer-term strategy needs to be put into place.

In April this year the Secretary of State announced a new five-year programme for the Valleys with a total expenditure budget of over a billion pounds. In general this is not new money but an effort to maximise existing resources. (Indeed the financial settlement for 1993/94 went extremely badly for the Welsh Counties. There are, for example, no new starts for highway projects in Mid Glamorgan.) Fourteen towns have been identified for urban development partnerships. He said:

The new programme will shift its emphasis to local partnerships and increasingly – important links with Europe. The new programme presents a fresh challenge, opens up new opportunities and sets a direction for economic and social regeneration. It should mark the beginning of a new age in the South Wales Valleys – an age of ambition realised, of growing assertiveness and prosperity, of a generation looking forward to better health and economic prospects than any generation before.

But when all is said and done, it appears that we do not have any mechanism – any organisation – for producing and implementing a comprehensive long-term regeneration strategy with clearly defined objectives, targets and programmes for the Valleys as a whole and which has emerged from the combined contributions of communities, local authorities and government. Mid Glamorgan County Council, the largest local authority in the Valleys, simply does not have the resources to take that initiative.

Land reclamation

On Friday 21 October 1966 the South Wales Echo carried the banner headline 'Coal tip engulfs school, village: 88 killed 65 still lost'. That terrible day in Aberfan provoked a dynamic programme of coal-tip reclamation. As so often happens in public policy, there had to be an awful crisis before positive action was taken. A survey of derelict land was carried out in England and Wales in 1972. Although Wales has only some 6 per cent of the population of England, it had 40 per cent of the dereliction of England. There was a further survey of derelict land in England in 1982 but curiously no such survey was carried out in Wales. The only explanation given to me (unofficially of course) was that, as there was such a lot of derelict land that could be tackled and only over 10 million pounds or so available annually, there was little point in finding out how much there was altogether. The original momentum for reclamation had dissipated.

By the mid-1980s it was clear that Wales was falling behind England and Scotland in its attack on dereliction. The Coalfields Communities Campaign in 1987 commissioned me to carry out a study when a Parliamentary question revealed that there was about as much derelict land in Wales in 1986 as there was in 1976. What was demonstrated by this study was that although Scotland had an objective of removing all significant colliery dereliction by 1991, there was no such clear programme in England or Wales. Indeed the study indicated that, at the current rates of reclamation, the dereliction of the nineteenth century would not be removed until well into the twenty-first century. The study recommended at least a doubling of annual resource allocation to around 25 to 30 million pounds a year. A survey of derelict land in Wales was carried out in 1988 but unfortunately the report has never been published. Happily, in the last few years this increase in resource allocation has been achieved. In January of this year the Secretary of State was able to say that they had set themselves the target of removing all visible industrial dereliction by the end of this decade. The Government and local authorities together are well on course to achieve it.

One man's industrial dereliction is another man's industrial archaeology. This industrial archaeology – charting as it does the shattering shift from an agricultural to an industrial society – is obviously precious in its own right, but it also has an added dimension: it becomes a key factor in the new patterns of tourism and leisure which themselves are major components of the new economy of Wales. I think it is fair to say that the Welsh Development Agency's land reclamation division seems to have been more conscious of this significance than some local authorities. The Agency gave support under its land reclamation powers to conservation (but unfortunately not restoration) of archaeologically significant items. They were considered under five headings:

- those capable of active commercial re-use;
- those with scope for amenity use;
- those capable of being actively promoted as 'commercial' monuments;
- dormant monuments; and
- sites complementing existing sites of archaeological interest.

Local authorities have, in more recent times, begun to be aware of the significance of industrial archaeology. But this has occurred at just the time when central government's squeeze upon their expenditure is at its most ferocious. In earlier times there was a tendency to look upon these artefacts of the past as ugly reminders of an age that was best forgotten. For those suffering the consequences of that age, that attitude is at least understandable.

Urban regeneration

Environmental and economic issues are becoming intertwined. The coalfield communities were born because the coal was there. Most economies' activities are no longer site-specific: they can move where they will over the face of the earth: low production costs, stable political systems and good access to markets are some of the location criteria. But there is also the criteria of environmental quality which is becoming crucial. An interesting aspect is the new opportunity for using the river frontage for enhancing the centres of many valley towns. During the nineteenth century many became little more than sewers so that there was a conscious turning of backs on the rivers. But now the rivers are becoming clean once more, giving exciting opportunities to revitalise the environments of the towns.

Cardiff, the regional centre as well as the capital city of the Principality, developed rapidly in the nineteenth century as the port and commercial focus of the Valleys' activities in coal, iron and steel production. So it too went into decline with its own chronic unemployment and environmental conditions. But in a strong pro-active way during the 1960s it developed a series of economic, environmental and transport policies, taking a long-term view to the year 2001 and with a clear 12-year development strategy. In 1987 the Cardiff Bay Development Corporation was formed to carry on the regeneration of South Cardiff and to create a maritime city of international quality. Again a comprehensive 15-year strategy was developed in economic, social and physical terms with a powerful focus on environmental quality. A partnership between the public and private sectors, with an additional budget of 500 million pounds of public investment and some 1,500 million pounds of private investment in this relatively small area of land, is propelling Cardiff into being one of the truly exciting cities of the twenty-first century, both in its quality of environment and in the quality of everyday life it offers its citizens. This, of course, has regional significance and regional impact.

Contemporaneously the Welsh Development Agency has created an Urban Development Division. In the last three years it has begun a series of exciting initiatives with individual towns. These initiatives have three characteristics:

- each is a joint venture with local authorities and other appropriate partners;
- the venture begins with a critical assessment of strengths, weaknesses, opportunities and threats; and
- medium-term economic development policies are set in a comprehensive framework of transport and traffic, environment, business, property development and marketing concepts.

These development activities are funded to the tune of 22 million pounds in 1992/3 and 34 million pounds in 1993/4. This is a worthwhile increase. It is of a similar order to that extra funding being given to Cardiff Bay Development Corporation to regenerate the southern part of Cardiff, yet this funding has to be devoted to regenerating all the towns of Wales – not simply the Valley towns. It is not that the level of resources being put into South Cardiff are too great, it is that the Urban Development Programme will need massive increases in the next two decades.

The European dimension

From the early 1970s the possibility of help from the European Community was recognised by local and central government in Wales.

By the late 1970s substantial funds were being fed into the South Wales Valleys for infrastructure and social purposes. Although the arguments about additionality went rumbling on, there is little doubt that the sheer existence of the structural funds made central government as well as local government much more aware of the need to pursue re-structuring policies. Signs were appearing all over the place saying 'This project has been assisted by the European Community.'

Wales has been fortunate that successive Secretaries of State have been pro-active in the European scene. Local government staff have been seconded to Brussels, the Welsh Development Agency has created an office in Brussels and the Agency is active in promoting links with other regions of Europe. Last year the Community's structural funds made a larger contribution to Wales than the combined Welsh Urban Programme and the Urban Development Programme.

The previous Secretary of State, David Hunt, has recently argued for a strategic framework of regeneration which can be accepted locally, nationally and by the European Commission. He also argued for a simplification and unification of the grant mechanisms that fund it. These are commendable objectives. The difficulties in the way of achieving them are formidable. A great amount of energy and resources are needed as inputs to resolving them.

Review

What lessons can we learn from this story of the coalfields of the South Wales Valleys? What initiatives should be taken? I think the first lesson is that the restructuring of the framework of a major coalfield community is a lengthy business. To think in terms of two or three decades is not being pessimistic but realistic. Long-term strategies need defining, monitoring and reviewing. South Glamorgan County Council is already carrying out a 'Cardiff 2020' study to begin to define priorities for action over the next 27 years. But it has established a tradition of long-term strategic thinking which has happily commanded the support of the major political parties.

The second is that these long-term strategies must not remain simply at the level of political policies: they need translating into economic, social and physical plans. The Cardiff Bay Development Corporation was established in April 1987. By the summer of 1988 it had commissioned and received a detailed planning study, showing how regeneration might be achieved over the next 15 years and the extra input of public section and private sector funding required to support it.

Since that time the over-view strategy has been broken down into more detailed area planning briefs. These briefs have been developed through an assiduous consultation process with the communities and businesses of the city, as well as the usual ones with statutory authorities and the other arms of local and central government. In this way all those who can, or should, or wish to, or must invest – whether in the public or private sectors – can see what part they can play. Naturally, such detailed plans need to be treated flexibly and be responsive to new ideas and proposals, but nevertheless they can provide sufficiently accurate guides to the scale of investment, the possible parties who might partake in the actions and the likely time-scale needed.

A comparison, for example, between the thorough, detailed and long-term strategies, planning briefs and public/private sector inputs which have been prepared for South Cardiff by Cardiff Bay Development Corporation and the fragmented, project-by-project, short-term urban regeneration proposals for much of the Valleys' coalfields, demonstrates that more integrated and long-term resource allocation to the regeneration of the region needs to be applied to the Valleys. Only in this way will we be able to develop a shared vision of the future with a realistic assessment of the commitment and the cost involved.

The problem is that no mechanism exists to achieve these thorough, detailed and long-term strategies and planning briefs for the Valleys as a whole. Yet they are crucial as a basis of claiming funding support, in achieving the high level of co-operative working required, in obtaining commitment to a shared vision. It is not a question of top-down or bottom-up planning. The kind of planning I am talking about is a reiterative process of both.

The restructuring of the coalfield communities is now recognised, especially by the RECHAR programme, as being a task that requires action and support at the European, state and local government levels. This implies policies, plans and programmes which are acceptable to each of these levels and a mechanism of funding equal to the task. Professor Stephen Fothergill has carried out pioneering work in forming assessments of the kind of funding required to meet the public sector input: my experience indicates that he has certainly not over-estimated them.

The third lesson is that, in preparing these policies and programmes, it is not sufficient to seek to bring the coalfield communities up to the level of today. They need to meet the requirements of tomorrow. This demands political and professional vision. I do not mean by 'vision' some form of dream or trance; I mean an imaginative insight; a statesman-like foresight; sagacity in planning. For example, our coalfield communities by their very nature are weak in research, design and innovative skills. Yet if Europe is to survive in the intensely competitive world of the next century – especially the economic challenges of the Pacific Rim including China – we need to find niche roles in which we are world class. Academia, business, government of all levels and the TECs, need to co-operate to perceive the ways ahead. The task is desperately important. We start from such a disadvantaged position. It worries me deeply that there are so few signs that, generally, we in Europe are yet aware of the scale of the growing economic might of the Far East. Yet the evidence is all around us. How do we rid ourselves of our complacency?

The fourth lesson is that environmental quality is not an optional bolt-on extra. It needs to occur naturally, permeating everything we do. This is for two prime reasons. The first is that the heart of public policy is to improve the sheer quality of everyday life of ordinary people. It is becoming increasingly evident that the environmental quality of our surroundings has a deep effect upon the quality of our lives. The second is that economic activities, and especially high added-value economic activities, are increasingly seeking locations offering a quality environment. The land reclamation policies of the Valleys are a remarkable story. But the environmental quality of our Valley towns leave much to be desired even by today's standards – even more by the standards of tomorrow. For the Secretary of State to promote the concept of simplified planning zones procedures as a remedy for the Valleys is a disappointing initiative. We need much more quality – not less – in our urban environments if we are to survive in the next century. The quality cannot be achieved by chance. It needs an intense kind of caring, an unremitting thoughtfulness, a special sort of love.

References

Fothergill, S. (1991), *The EC's Rechar Programme: an adequate response to the problems of coal areas?*

Parkinson, E. (1987), *Local Reclamation in Wales*: Study commissioned by the Coalfield Communities Campaign.

Mid-Glamorgan County Council (February 1993), *The Future of the Valleys: Proposals for Action*.

Welsh Development Agency (1991/92), *Report and Accounts*.

Welsh Office (1967), *Wales the Way Ahead*.

PART VI

EDUCATION AND TRAINING: RESKILLING THE WORKFORCE

15

TRAINING AS A MEANS OF PROMOTING THE RESTRUCTURING OF THE COAL REGIONS

DR HEIKO BEENKEN

Introduction

Miners throughout Europe fear for their livelihoods. Many have already lost their jobs and are looking for new employment. This is no easy task, in view of the European employment crisis and economic recession, and is one which is causing great concern to mineworkers' families and putting them in financial difficulties. The reasons for this trend are well known and need not be repeated here.

The Federal Republic of Germany and North-Rhine Westphalia in particular, the Federal Land in which 86 per cent of Germany's hard coal is produced, has been severely hit by this crisis in mining. Ruhrkohle AG, based in Essen, which owns almost all of the pits still working in North-Rhine Westphalia, announced back in 1992 that some 36,000 jobs would have to go in mining by the year 2005. However, the recent development in the European steel industry – the obligation to reduce excess capacity – has increased the pressure for even quicker and larger-scale staff cuts in North-Rhine Westphalia too. Staff adjustment measures are having to be given priority; further serious decisions can be expected.

This development and the current situation is presenting North-Rhine Westphalia's employment policy with specific challenges. In the context of this paper, I would like to outline:

- the employment policy aims being pursued by North-Rhine Westphalia; and
- the social policy instruments generally available to prevent unemployment among mineworkers.

I would then like to focus on one employment policy instrument, the promotion of training schemes, and then to explain:

- the various types of training schemes;
- their concrete organisation; and
- the financing of the schemes.

I would like to conclude with a report on the results achieved to date and prospects for the future.

The aims of North-Rhine Westphalia's employment policy

In the Federal Republic of Germany employment policy is a matter for Federal government. It uses particular Federal institutions (employment offices) within the framework of a specific Federal law (the Employment Promotion Law) to promote employment policy schemes. Because in the past, and this is currently the case too, continuous reduction in the scope of central government employment policy resulted in rising unemployment, North-Rhine Westphalia developed its own employment policy instruments. These specific schemes are financed from the Land's own funds and grants received from the European Social Fund. The main aim of the Land's employment policy programmes is:

- to thwart unemployment by means of preventive employment policy measures; and
- in particular, to reintegrate the long-term unemployed into working life.

When it comes to mining, various agreements in the past between Federal Government, the Lands, trade unions and mining companies have actually made it possible, to a large extent, to prevent any miner from being left to his own devices. However, the resolutions passed at the 1991 Coal Round, whereby over 36,000 jobs are to be cut in North-Rhine Westphalia alone, have presented special challenges to all those responsible. They will require the use of all available means, including job creation schemes in coal-mining districts and measures aimed at preventing unemployment. There are several social policy instruments designed to prevent unemployment.

Social policy instruments to prevent unemployment among mineworkers

The following social policy instruments are available: the movement of employees to other pits; early retirement; permanent short-time working; training schemes. The first instrument, involving the movement of employees to other pits, does not reduce the total labourforce. There are now scarcely any operations with the capacity to take on extra workers. In some cases younger workers from other pits can replace older workers accepting early retirement.

The second instrument, early retirement, is of major importance in German mining. Employees can take early retirement at either 55 (surface workers) or 50 (underground workers). They receive adjustment money until they reach formal retirement age; this is financed by the Federal Government and the Lands. In 1992 there were 14,900 former miners in early retirement; by 1993 there were 16,800. It was planned that a further 20,000 miners should be given early retirement by the year 2000. It is impossible to predict whether these figures will be reached, in view of the relatively low average age among mine workers (31). This instrument is likely to become rather less important as time goes by.

The third instrument, permanent short-time working, reduces the burden of wage payments on mining companies. Miners receive 68 per cent or 63 per cent of their last net wage from the state employment authorities and do not work in the company during phases of short-time working. This instrument does not reduce the total workforce either. It does, however, lead to a reduction in actual coal output. Permanent short-time working may last no longer than two years in the Federal Republic of Germany. In some cases, it immediately precedes early retirement. However, due to the relatively high level of fixed costs in mining, which have to be met even without production, this instrument is only used sparingly.

The fourth instrument, training schemes, has become increasingly important since 1991, as the other instruments mentioned earlier are not sufficient to make the staff cuts in mining socially tolerable. The way in which these are organised will be explained in greater detail next.

To complete the general outline of social policy instruments, I would also point out that mining employees can, of course, cross directly into other fields of activity without the aforementioned instruments, something which is also practised with the support of employment agencies.

In some cases, wage subsidies are also paid to new employers, provided they employ former mine workers.

Types of training

A distinction can be drawn in terms of timing between part-time training schemes, in which training is given in addition to professional employment, and full-time training schemes, in which training is given during normal working hours. Depending on the mineworker's status, a distinction must be drawn between training schemes after the worker has been made unemployed and training schemes while the worker is still employed in mining. In North-Rhine Westphalia, miners are trained for employment outside mining, even while they are still employed in mining. Employment only ends once the training has been completed. This conforms to the principle applied to date, whereby employees in the mining industry are not made redundant for operational reasons. Depending on the aim of the training, a distinction must be made between so-called adjustment training schemes, where further knowledge is gained, in addition to the occupation previously learnt; and so-called retraining schemes, in which a completely new occupation is learnt.

Organisation of the training schemes

Training schemes for miners threatened with unemployment are developed by the mining companies themselves in collaboration with the local economic and training institutions. The planned schemes are discussed and evaluated on site in the individual coal-mining districts at so-called regional conferences. There are currently a total of 12 such regional conferences in North-Rhine Westphalia, where local government, chambers of industry and commerce, employers' associations, trade unions and welfare establishments are represented. These conferences also develop their own proposals and have a crucial effect on subsequent sponsorship decisions. Once the training schemes have been developed, they are presented to the mineworkers at the pits. After they have registered for individual schemes, employees enter into a so-called training contract, which provides for continued payment of their agreed salary until the training scheme has been completed, followed by their departure from mining.

The training schemes are run by the mining industry's own educational establishments and other external educational establishments, working in close collaboration with companies looking for workers in the regions concerned. In the early planning stages of such schemes, various contacts are made with those branches of industry looking for workers; in some cases, the Ministry for Employment, Health and Welfare and the Ministry for the Economy will also mediate. The transitional process from the training scheme to new employment begins as soon as the schemes start and is supported well before the scheme ends by the state employment authorities.

Training scheme finance

The training schemes can be financed by the state employment authorities (maintenance grants for participants during the scheme); by labour market programmes co-financed by the EC and the Land (Target 2, RECHAR – scheme costs are subsidised by up to 80 per cent, with maintenance grants for participants in some cases too); and by the mining industry itself. In addition, there are plans to use funds from the EC Social Fund, which are to be made available for such purposes.

The cost of the training schemes is between five and 13 DM an hour for full-time schemes, depending on the type of scheme involved. In addition to this, there are the maintenance grants for participants during the schemes, which are equivalent to between 68 per cent and 63 per cent of the last net wage.

Attempts are being made to finance the schemes jointly using all the aforementioned sources of finance. This means that applications must be made to the different institutions responsible in each case. Guaranteeing the total financing requires a great deal of co-ordination. A working group has been formed to deal with this. It includes representatives from Ruhrkohle AG, the mining unions, mining educational institutions and employees from the Ministry for Employment, Health and Welfare, who liaise with the state employment authorities.

Results to date

The financing of 1,000 places on training schemes for mining employees is currently guaranteed. The first larger-scale schemes started on 1 April this year; additional ones will follow on 1 July and 1 September of this year. The mining companies have already indicated that they envisage the need for the creation of a thousand more such training places, in view of the crisis in the steel industry. The public financing of such schemes will obviously be a problem, since the EC co-financed pro-grammes are to finish at the end of this year. North-Rhine Westphalia is therefore pressing for both the Target 2 and RECHAR programmes to be continued from 1994. The EC Commission has yet to make a final announcement on these requests. We can only hope that it does not close its eyes to them, in view of the problems in the coal-mining districts.

16

THE IMPORTANCE OF FURTHER EDUCATION DURING STRUCTURAL CHANGE: A CASE STUDY BASED ON THE TOWN OF BERGKAMEN

ROLAND SCHÄFER

The economic importance of training and further education

Modern industrial society requires a lifetime of learning. Reasons for this are varied: change in products, processes and consumer interests, technological progress, especially electronic data processing, development of markets e.g. the unification of the European market, long-term trends e.g. the growth in the service sector, and, last but not least, the pressure of competition on the world market, especially by the rising industries of South East Asia and Eastern Europe. For industry these developments bring the demand for better qualified employees, even if fewer in number. For the employee, training and further education means better job procurement and a higher grade of job security.

One should not forget that the basis for the ability of life-long learning is laid by school, university education and vocational training. This basis should have priority. Further education is just what the term implies: an additional and supplementary form of training. Due to the above-mentioned world-wide developments, further education is a vital element for every industrial society. This holds even more for the old industrial regions of coal and steel. In these regions, great numbers of employees are released by the decline of the traditional enterprises. Employees, who are often highly qualified for their former jobs as miners or steelworkers, now have the need for additional training and further education to cope with the demands of jobs in new and different industries.

Bergkamen – portrait of a town undergoing structural change

One example of a town undergoing severe structural change is the town of Bergkamen. Bergkamen is a town with 50,000 inhabitants, geographically situated in the North-Eastern Ruhr area, between the cities of Dortmund and Hamm, politically, part of the district of Unna, in North-Rhine Westphalia. The town administration consists of 530 employees. In 1993 the town had a revenue budget of 161 million DM and a capital budget of 47 million DM.

The municipal infrastructure is more or less complete: schools of all types are present, as well as facilities for youth, sports, culture and recreation. Bergkamen has an excellent traffic network, with two major motorways in the municipal area, a shipping canal, railroads and the commercial airport of Dortmund nearby.

The rate of unemployment in 1993 was about 10 per cent. The economy was and is dominated by coal-mining and chemical industries. In 1991 there were, all in all, about 22,000 jobs in Bergkamen, of which 10,000 were in two coal pits. During 1992 to 1994 the two pits were merged and the number of jobs reduced to 4,000, i.e. 6,000 jobs were lost. In the neighbouring towns coal pits and other mining industry facilities have been closed as well. More than 15,000 jobs were lost in the whole region over the last four years.

Further education – one aspect of a strategy for coping with structural change

One has to keep in mind that further education is only one factor in an integrated and long-term strategy for coping with structural change. The first step in working out such a strategy is analysing the inherent strengths and weaknesses of the municipality. The next steps are formulating aims and finding appropriate measures to achieve them. This process cannot be confined to the town alone; the whole region has to be included. In the same way it is absolutely necessary to seek the participation of the public and of all relevant interest groups in the region, i.e. municipal and state authorities of all administrative levels, industry, labour unions, science, political parties, professional associations, chamber of industry and commerce, churches, social groups and so on.

On the initiative of the government of North-Rhine Westphalia, regional conferences were formed in 1992 consisting of representatives of the above-mentioned groups. These conferences meet regularly to work out a regional development strategy. In the end, the town council, being politically responsible and having the exclusive right of town planning, will have to decide which direction to take.

In Bergkamen it was decided to try to remain an industrial town. This means, besides safeguarding the remaining jobs, creating new ones by attracting small and medium-sized industrial enterprises outside mining. The necessary measures to achieve this goal were decided to be: planning allocation, reconditioning and development of industrial areas, support of technology transfer and further education, increasing the site's general appeal for investors in the areas of housing development, urban construction, local government infrastructure and environmental protection.

Further education was included as one of the focal points because of its general importance for coping with structural change and because of the already existing private and local government further education establishments in Bergkamen and in the district of Unna.

Private further education establishments in Bergkamen

Throughout the region, i.e. the district of Unna and the cities of Dortmund and Hamm, there are more than 200 further education establishments. In Bergkamen there are five minor facilities and two major ones. Minor facilities are: Bergkamen School of Physiotherapy, Bergkamen Seminar for old people's nurses, Bergkamen Women's Workshop, Evangelical Adult Education Institute, Evangelical Family Education Centre, Catholic Youth Training Institute. The major private establishments in Bergkamen are the Westphalia Technical Control Board Academy (TUV Akademie) and the DMT Vocational Mining School East (DMT-Bergberufsschule Ost).

The Westphalia Technical Control Board Academy, Bergkamen, is sponsored by the Rhineland Westphalia Technical Control Board (RW TUV). Its staff consists of 25 full-time employees and approximately 400 part-time lecturers. Target groups are skilled workers and management personnel from manufacturing and service industries and local authorities. On offer are more than 400 full-time seminars from one to five days duration, approximately 20 part-time courses ranging from six months to two years and in-house company training. The seminars and courses cover production technology, safety and quality, logistics, advertising, sales, management and executive techniques, personnel and welfare, foreign languages, data processing, corporate environmental protection and legal questions. The instruction is financed by subscriptions of the participants or the respective company. For the unemployed or employees in danger of losing their jobs, special courses are offered, if public financial support is available. This financial support consists mainly of retraining benefits from the Employment Office under the Employment Promotion Law (AFG), grants from Land and EC funds (especially the European Social Fund and RECHAR). In 1993, for example, 1,300 employees from the mining supply industry of the region, whose jobs are gravely endangered by the decline of mining, underwent special courses to qualify for jobs in other branches. This project was supported from EC and Land funds with a grant of 1.6 million DM. The DMT Vocational Mining School, Bergkamen, is sponsored by the Deutsche Montan Technologie – Gesellschaft für Lehre und Bildung mbH, a Ruhrkohle subsidiary. Its staff is made up of 33 full-time teachers, 12 part-time lecturers and one member of administrative staff. The school has a capacity for up to a thousand pupils, with 42 classrooms, five laboratories and workshops, three lecture theatres and a refectory. Target groups are mining trainees and Ruhrkohle employees. The school contains three branches: a Vocational Mining School, a Higher Technical School of Technology which offers qualifications for mining technicians and a Technical Mining School of Technology offering qualifications in mining engineering.

In 1994 DMT and RW TUV formed a new holding company (CUBIS) to combine the capacities of both establishments for general advanced vocational training and further education in the technical field. They plan to offer seminars and courses in technical environmental protection, hydraulics, pneumatics, explosive technology, laser technology, systems electronics and control technology.

The backing from the town of Bergkamen for private further education establishments in Bergkamen is only in one case of a financial nature: the Bergkamen Women's Workshop gets a staffing cost subsidy of 20,000 deutschmarks a year. In some cases, favourable terms for the use of municipal premises are granted. The main backing is in the form of assistance when dealing with authorities, i.e. making contacts and giving administrative help.

Local government establishments

Beside the careers guidance offered by the Employment Office and the Chamber of Industry and Commerce, there are three local government establishments in Bergkamen, two with advisory functions and the Bergkamen Adult Night School. The BBG, an establishment which no longer exists, may serve as an example of how not to run things.

The first advisory establishment is the Unna District Further Education Trust (WKU-Stiftung). It is sponsored by the Unna District Economic Development Company (which, in turn, is sponsored by the district and 10 local authorities), the Unna district, the Chamber of Industry and Commerce, the Chamber of Crafts and the Employment Office. The trust capital is currently one million DM. The staff consists of two part-time directors and an advisory staff of two specialists. Services offered are advice and information on matters of further education for everyone, companies and further education agencies, co-ordination of further education provision and analysis of future further education needs. The trust publishes a further education yearbook, via an electronic database which contains all available training opportunities in the region. Costs are 180,000 DM a year, which are covered by a yearly lump sum of 100,000 DM from the Land and the interest on the trust capital. The future of the trust depends on continued financial backing from the Land or an increase in trust capital.

The second advisory establishment in Bergkamen is the Woman and Career Regional Office (ZEFF). It is sponsored by the town of Bergkamen and three neighbourhood local authorities. The staff consists of two advisory staff, one administrator and 10 part-time lecturers. The target group is girls and women, particularly those returning to work after having families. The office offers help with contacts at the Employment Office, companies and further education establishments. Careers guidance and training in how to apply for jobs are also offered. The costs are 360,000 DM a year, 70 per cent of which is supported by Land funds and the remainder is provided by the sponsors. Bergkamen's share is 28,000 DM a year. The future of the office depends on continued financial support. If the funding by the Land should end, the local authorities probably won't be able to take the whole financial burden.

The Bergkamen Adult Night School is another of Bergkamen's institutions. The staff consists of two full-time teachers, six administrative staff and about 140 part-time lecturers. The target group is all adults.

The school offers lectures and evening classes with about 300 events in total with approximately 11,000 lessons. The main contents are secondary school leaving examination, higher technical school leaving examination, commercial practice, office technology, data processing and foreign languages. The costs are 1.36 million DM a year. This is covered by 358,000 DM from Land allocation, 158,000 DM from participant subscriptions and a residual amount of 844,000 DM from the town's budget. The tight budget will force the town, in future to raise the participant subscriptions and take rationalisation measures.

An institution which was only in existence from 1984 to 1990 was The Bergkamen Berufsausbildungsgesellschaft mbH, otherwise known as the BBG, which had the aim of combatting youth unemployment. It offered vocational and prevocational training for young people at four training centres with workshops and three commercial training companies. There were 23 craft and office occupations courses ranging from locksmith to screen printing. Approximately 1,900 young people completed their vocational training, 90 per cent of whom found a job using their qualifications or went on to higher education. In addition, jobs for more than 170 unemployed craftsmen, teachers, educational sociologists and administrative staff were created.

The financing of the BBG was based on the systematic use of all financial support from federal government and the Land. Additional funding by the sponsors was not planned; residual finance was expected from the sale of products and services produced during training. In 1988 financial irregularities and problems were discovered in the company. It gradually became clear that the financial system did not work as well as it should. There was a considerable deficit from the beginning but this was concealed by cross financing, i.e. existing deficits were disguised by using funds for new projects. There was also misappropriation and falsification of funds by the director and the book-keepers.

The town and district councils rejected the idea of allowing the company to go bankrupt. Bergkamen and the district of Unna covered the deficit so that the young people could complete their training and the staff were given help to find new jobs. The main reasons for the BBG's unexpected deficit were: an unrealistic financial plan, an overworked and criminal director, a gullible supervisory board and inadequate financial control by the town and district administration. The total costs of the BBG amounted to 94.28 million DM, of which 75.14 million were covered by the financial support of the Federal government and the

Land. The town of Bergkamen and the district of Unna had to bear the actual deficit of 19.14 million DM.

Conclusion

Modern industrial society requires a lifetime of learning. Advanced vocational training and further education are important factors in coping with structural change. Further education has to form an integrated part of an extensive and complete strategy for regional and town development. Further education for the unemployed largely depends on public finance. Local government establishments require professional management and effective financial control.

17

COMMUNITY EDUCATION, TRAINING AND LIFELONG LEARNING: NEW PARTNERSHIPS AND INITIATIVES FOR SOUTH WALES

PROFESSOR HYWEL FRANCIS

Introduction

In writing this chapter I have very strong evocative memories of the 1984–85 Miners' Strike. These memories have contributed to my own lifelong learning. I hope that you will bear with me while I indulge myself in these reflections. I also hope that you will agree that they are not unconnected with some of the key themes which concern us.

1985: The return to work of Cynheidre (a pit in the Gwendraeth Valley in the Western Anthracite Coalfield). The words that come to mind are community, lifelong learning, progress, union solidarity, development, citizenship, dignity, work. The men and women of the coalfields talked the language of culture, community and democracy.

In reflecting on these memories I believe that there are four particular themes which are relevant to this book as a whole: namely, lifelong learning, partnership (in the old fashioned meaning of unity and solidarity), community and a 'wider vision'.

As a young friend of mine from Aberfan once said, 'We have too many missionaries in the valleys: what we need are visionaries but visionaries whose feet are firmly on the ground.' Too often I believe we indulge in recreating a past which never existed (or only partially existed). And that is my starting point. I will begin with some of the problems which we now face, problems beyond the obvious economic ones, problems which should start from the position of asserting the idea that we are citizens with rights not subjects with duties to the State or Crown. Some of this 'new times' thinking began in the Neath Conference of September 1992 which raised questions more about the future than the past. It is in this spirit that this chapter is written.

Four basic problems:

Ideological

What do we mean by 'community'? It is a word that is so widely used by so many in different circumstances: community programmes; community care; community development; community education. It has been suggested that sometimes the word 'community' can act as a smoke screen to fudge key issues about power, accountability and resource allocation. The adult educator Raymond Williams believed that it was a notoriously 'slippery' word. He said 'community' has become a ubiquitous label, partly because it affords those who peddle it the luxury of not being pinned down too precisely. How does it relate to social class, gender, ethnic minorities? Whose community are we talking about? Community education should therefore be about partnership and solidarity rather than paternalism and manipulation.

A recent study of Adult Continuing Education by the Open University highlighted three major areas of debate on Adult Continuing Education which are worthy of mention:

- Who has access to what forms of learning?
- What is the relationship between education, training and the economy?
- What is the role of community-based adult learning in changing power relations and improving quality of life?

It might in future be better to talk about popular education and lifelong learning and to define these in this way:

. . . starting from the problems, experiences and social position of excluded majorities, from the position of the working people, women and black people. It means recognising the elements of realism in popular attitudes to schooling, including the rejection of schooling . . . It means working up these lived experiences and insights until they fashion a real alternative. (Edwards *et al.*, 1992, p.194)

Cultural

The collapse or decline of our basic industries in South West Wales –
coal, steel, tinplate, and to an extent agriculture and indeed defence –
has had a consequent effect on the culture of our region. But we still live
very often in a time warp, believing that we still inhabit that world which
we have lost. I am not at all sure whether these Miners' Welfare Halls
can entirely be regenerated and rejuvenated. Many of them now have a
paternalistic air to them and lack the sense of participatory democracy
that perhaps they once had.

The major regional organisations, especially the voluntary ones which
were so vital and dynamic, have now entirely disappeared. The two most
important – the co-operative movement and the miners union – were
extremely important in providing a whole range of voluntary services
and educational opportunities. Local education authorities, for so long
over many generations the major providers of adult education, are, as a
consequence of recent legislation, no longer key players. It is important
to recognise that these bodies were not only providers but also were
engaged in an important democratic partnership with the communities
which they served.

Inherent in that world we have lost was a regional unity. There was a
sense of strong democratic partnership, indeed solidarity, throughout our
region, a region which was often referred to simply as 'the anthracite'. It
was a region which was rich in its voluntary organisations but also rich in
its Welsh cultural diversity. A region which, even today, is the only
urban/industrial region in Wales which is still predominantly Welsh-speak-
ing. That, in itself, poses problems and challenges to lifelong learning.

It may well be that we will need to start afresh and leave some or all
of this baggage behind. Something happened to Miners' Welfare Halls,
which were the focus of many of these movements, in the 1950s and
1960s when their libraries gave way to drinking clubs. And while we
would wish to take forward some of the best of that period it may well
be that we need to think more imaginatively about the new initiatives
and the new challenges before us.

All this also leads me to reflect on the fact that we are a multi-lingual
and multi-cultural society, so that our rights as citizens should be based
not on language of ethnicity but on justice and equality for all.

Educational and training 'achievements'

There has been a tendency over many generations to be rather uncritical
of our Welsh educational system. This often in more recent times has

masked a very serious problem, particularly among young people. There is undoubtedly, in the present economic climate, an air of declining aspirations, of low self-esteem, especially among those who feel socially excluded from an educational system and economic system which appear to have failed them. The recent report by the Institute of Welsh Affairs entitled *Wales 2010* highlights this. The percentage of young people attaining the equivalent of NVQ level 3 in 1991 shows a shocking disparity between England (29 per cent), Scotland (42 per cent) and Wales (25.5 per cent). It is undoubtedly the case that the more integrated Scottish educational system is at many levels further down the road of transforming itself for the needs of the modern period, particularly in vocational education, than is Wales.

The regional identity

A recent study by Adam Price for the BBC, entitled *The Forgotten Valleys,* highlights a general feeling in the Amman and Gwendraeth valleys that they are being neglected because they have 'little political clout' compared with the valleys to the east. Beyond this one could say that there is a general feeling as a consequence that they are living, even more than other declining areas, in a dependency culture simply because they have a lack of democratic leverage. A lack of employment opportunities, a deteriorating environment (the expansion of open-cast mining) and a decline in many other measures of the quality of life, result in a general feeling of hopelessness on the part of this region.

Posing solutions

It is often reassuring to reflect on situations elsewhere. The description below is not of Britain but Appalachia in the USA:

Since miners, by achieving living wages, safer working conditions, pensions and better health care for their families in communities having inspired the efforts of others, the destruction of their gains is doubly tragic.... Yet crisis forces change and change brings the possibility for effective new responses to the many problems confronting coal mine families...to support their families an increasing number of miners wives have joined the labour force. This fosters adjustments of the relationships between husbands and wives which may lead to a new solidarity for the surival of their families in their communities.... The type of future that miners and their families are struggling to create however remains largely undefined. Since the coal industry has so visibly shaped their region for its purposes they have great difficulty in imagining a future to fight for if their coal is no longer in demand. (Yarrow, 1990, p.38/45)

Implied in this quotation is a feeling that we are in a new situation – a post-coal era, maybe even a post-industrial era. But also it is an acknowledgement that much has been learnt individually and collectively.

If we for a moment imagine we have a clean slate, what is the new context for education and training for Gwendraeth Valley in Wales in Europe in 1994? The following are positive features of our new times. Firstly, all Further Education and Higher Education institutions in South West Wales are committed to widening participation as are employers, trade unions, local authorities and a range of key agencies such as the Welsh Development Agencies and West Wales TEC. Secondly, The Welsh Office is developing a strategic framework on skills and enterprise. It has three objectives:

- identifying clear priorities which reflect Welsh needs;
- creating effective teamwork between different agencies;
- adopting programmes and funding arrangements to underpin this work.

Beyond all of this the Welsh Funding Councils led by Professor John Andrews also have an enabling outlook.

Thirdly, Hywel Ceri Jones of the European Commission, in a recent lecture at University College Swansea entitled Education in a Changing Europe, emphasised the severe skills mismatch problems throughout Europe:

The European education and training systems are not providing the minimum competencies and qualifications for the vast majority of young people to get into the labour force – this is the direct route to social exclusion. Let us not forget that in the EC as a whole we have twelve million unemployed of which more than four million are young people under the age of 25. The training systems are not fast enough in their response to new skills needs resulting from technological change in sharpening economic competition – hence a fundamental mismatch problem which is typically European. Social policies have remained in a passive 'income maintenance posture' despite the fact that loss of skill in adult life is now the main threat of social security, leading directly into long-term unemployment and dependency on state income. (Jones, 1992, p.9)

He went on to cite what the employers group in Brussels recently produced for the Commission as a list of the skills which young people should acquire in compulsory schooling and develop more fully in vocational training:

. . . commitment to quality; ability to communicate effectively; knowledge of one or more languages; the desire to use a developed technology; a general grasp of the importance of the economic environment in which firms function; problem solving capacity; willingness to adapt to change; ability to work in a team and relate to others; and an understanding of economics and the labour market. (Jones, 1992, p.16)

There is therefore a general recognition at regional, Welsh and European levels that education and training and lifelong learning is central to our economic and personal development. But how do we get into a position of effecting change in the medium term? There would be many among us who would say that the whole question of social and political exclusion is one which is debilitating us in Wales, that some kind of democratic accountability needs to be injected into the situation in order for us to get some sense of ownership. The Valleys Initiative talked of a 'partnership with the people'. What we need is to develop and democratise that relationship where a public dialogue acknowledges the centrality of openness, accountability and regional/national (Welsh) unity and coherence.

Programme for action?

In the British context this situation is currently very difficult: the previous stability and order provided by Local Education Authorities can possibly degenerate into very considerable chaos as Further Education Colleges are put in a position of competing against one another. It is therefore encouraging to read the NIACE (National Institute for Adult Continuing Education) discussion document entitled *The Learning Imperative* (1993) on national education targets and adult learners, which puts the onus on everyone to accept their responsibilities and to try to work in a collective manner. Its general recommendations are ones which everyone would endorse but often need to be re-stated. In particular, the NIACE document identifies the need to develop additional educational opportunities for people temporarily outside the active workforce; the need to promote the development of more accessible provision of education and training opportunities for adults with regard to both location and timing; and, perhaps most important of all, the need to take steps to develop practical learner-support services to overcome barriers to adults participation in education and training. This last point about guidance and counselling is one which should inform all our activities and it is very encouraging that the Welsh Office has taken a lead in this respect with the funding of the Valley Women's Roadshow.

The NIACE document identifies a range of short-term action points. One, in relation to the Government which needs seriously addressing, is the reconsideration of provisions in the Trade Union Reform and Employments Right Bill and any subsequent legislation which would limit the scope of access to advice on careers, education and training for adults. The document goes on to identify responsibilities for TECs, employers, funders of education and training, providers of education and training, broadcasters, and interestingly enough, trade unions themselves. Trade unions are asked to identify education and training as an essential part of reward packages when negotiating with employers and to promote and take practical steps with management to introduce employee development initiatives which provide a learning entitlement for all employees.

Beyond all of this, of course, the most critical issue throughout the whole of Britain, which makes widening participation and lifelong learning difficult, is the lack of proper funding for part-time study and the likelihood that the situation will deteriorate rather than improve over time.

In a Welsh context, there is a growing realisation that there is a need for coherent national and regional voices; that fragmentation is not desirable. A number of national bodies are now emerging in the new environment of funding being located in Wales rather than Whitehall. For example, Fforwm, the body representing the Principals in Further Education Colleges, and UCACE Cymru, the professional body representing continuing education in Universities, are two examples of the way in which this coherence is emerging. The Institute of Welsh Affairs has also flagged up the need for an all Wales development body for education.

In a regional context there must be a recognition that a variety of agencies must work in collaboration. I was struck by Charlotte Rees' recent study for the WDA entitled *Collaborative Training for Quality in the Automotive Industry* (May 1993). This was a study in South East Wales where all the TECs and employers were working together. There were some interesting proposals but the most striking feature of this study was the general desire for collaboration between employers who wish to pool their training resources. One omission identified by the study was the fact that Further Education and Higher Education Institutions and the trade unions were not part of this emerging partnership. There must be a recognition that Further Education and Higher Education and trade unions must play a decisive partnership role in South West Wales. The Wales 2010 study quite rightly emphasises that one of the crucial features of success in such regions as Baden

Wurttemberg is a very powerful network of businesses, institutions of education, trade unions and other bodies. Clearly, we are a long way from that in South West Wales and in Wales as a whole but there are some encouraging developments such as the new business partnership in the Amman and Gwendraeth Valley funded by the Welsh Development Agency. I believe, however, that partnership must be broadened and democratised to include a wider variety of community based organisations, as well as the local institutions of Further and Higher Education.

The university's role

The University College Swansea, through its Department of Adult Continuing Education, has played an enabling role in the recent past in setting up such networks as the South West Wales Open College and Access Consortium and the Valleys Initiative for Adult Education. It prides itself in its strategic role and wishes to broaden that role in partnership with the kinds of bodies I have been discussing. I am particularly intrigued by the possibility of a University/Community Axis involving all other bodies which see the importance of education and training as a central part of our future survival and development. We would wish to develop a European/ International Research and Development Centre in education, training and cultural development which would:

a) look at ways of improving participation of non-traditional groups (this Group in the Gwendraeth and the Valleys' Women's Roadshow are examples of this);
b) explore new FE/HE and community education initiatives for the benefit of the region which are negotiated with local communities as we are currently doing in several localities;
c) explore the link between popular culture and lifelong learning in order to advocate greater participation in education and training;
d) engage in 'respectful intervention' in areas where provision of Higher Education is weak or non-existent, such as Pembrokeshire – the Welsh Funding Councils appear to be supportive of such developments;
e) most important of all, achieve a greater regional identity based on the twin pillars of democratic partnerships and the centrality of flexible and negotiable quality education and training, underpinned by a new research and development culture; and
f) assert the concept of citizenship – that people have educational and cultural rights – as a universal concept.

One major development on this front has been the Community University of the Valleys which is the culmination of six years of patient work in one valley community, Banwen in the Dulais Valley. This is a unique venture in taking the university back to the communities which created it in the first place. The intention is to provide opportunities to groups of people who would for a variety of reasons be unable to attend at a university campus on a full-time basis. These include women returners, the long-term unemployed, the disabled and those with a range of caring responsibilities. But this concept is neither new nor unique (see a similar initiative in Nicaragua).

Universities have national and international missions; but they also have responsibilities to their localities and in re-focusing ourselves on the problems and challenges of South West Wales we will inevitably enrich those communities as they will enrich us. That would be a truly learning culture that will have grown out of a genuine democratic partnership between Y Brifysgol a'r Werin (The University and the People). And that would be the best university definition of 'Serving the Community' and linking community-based learning to the universal idea of our citizens' charter.

References

Edwards, R., Sieminsky, S. and Zeldin, D. (eds) (1992), *Adult Learners, Education and Training*, Open University.

Price, A. (1992), '*The Forgotten Valleys*', BBC [Institute of Welsh Affairs, (1992), *Wales 2010*].

Jones, H.C. (1992), *Education in a Changing Europe*, University College Swansea.

National Institute for Adult Continuing Education (1993), *The Learning Imperative*.

Rees, C. (1993), *Collaborative Training for Quality in the Automotive Industry*, Welsh Development Agency.

Yarrow, M. (1990), 'How miners' families understand the Crises of Coal', in J. Gaventa, B. E. Smith, A. Willingham (eds), *Communities in Economic Crises, Appalachia and the South*, Temple University Press, Philadelphia.

PART VII
AFTERWORD

18

SOME COMPARATIVE REFLECTIONS

CHAS CRITCHER

Introduction

Those of us who listened to the papers originally given over two days in Sheffield in June 1993, on which the chapters of this book are based, learnt a lot about how pit closures have been handled in Germany and Britain. Though there were clearly some factors in common, the overwhelming impression, from the first paper to the last, was one of difference. Since one of the declared objectives of the conference was to exchange experience, some reflection on the reasons for and nature of those differences seems apposite.

I should make it clear that I am no expert on coalfields in Britain or Germany. It fell to me to open and close the conference largely because I am head of the research centre in Sheffield Hallam University where the indefatigable Dave Waddington has led our research into mining communities over the last 10 years. I thus came to the conference as an interested party with some limited knowledge of the British situation and only the haziest of that in Germany. The reflections that follow may therefore seem obvious to those already able to make the comparison; for me, it was extremely revealing to see how apparently the same problem, of a contracting coal industry, could be handled so differently in Germany, specifically North-Rhine Westphalia, compared to what has happened in Yorkshire and elsewhere in the British coalfields.

The basic difference, it seemed to me, could be summed up in one word: planning. In Germany, there was an attempt to plan everything: the annual production of coal, the closure of pits, the training of miners, the reclamation of land, the regeneration of the local economy.

Suspending for the moment the thorny question of just how successful such planning has been in practice, the most striking thing was the clear commitment to planning as a principle. In Britain, by contrast, the only extant plan seems to have been the long-term intention to privatise the remnants of the coal industry. Symptomatic of the improvised and incremental nature of the British trajectory was the reaction of Secretary of State Michael Heseltine to widespread political opposition to the wave of proposed pit closures originally announced in October 1992. Seventy-five million pounds of special aid was subsequently made available; it was to be channelled through the Training and Enterprise Councils (by no stretch of the imagination representative bodies); they were given two weeks to submit proposals to the Government. In such a context, any kind of serious planning is not viable. It is the contrast between this and the complex web of co-ordination and consultative bodies directly responsible for handling pit closures in North-Rhine Westphalia which this afterword seeks to explore.

The common context: the market for coal

The driving factor behind pit closures, which appears initially to be the same in both locations, is the change over the last 20 years in the economic market for coal. Though in fact, as the introductory chapters by Waddington and Parry and Schubert and Bräutigam show, pit closures had already started to happen in the 1960s, the market for coal was at that time relatively stable. The key factors on the supply side were that coal remained the most available and cheapest form of energy supply. On the demand side, the major markets for coal were in the steel industry and the electricity supply industry. In Britain, such stability was enhanced by the fact that these were all nationalised industries but, even in Germany, the market had a built-in equilibrium.

However, these were essentially national markets; the most significant changes were to stem from the increasingly international nature of the energy market. Other forms of energy, especially gas and oil, began to compete directly with coal and hence to shrink its market. Even within the coal market, cheaper-priced coal became available from all over the world: from Poland, Columbia and South Africa. Simultaneously the world demand for steel began to fall and there was fierce competition from cheaper-priced steel, especially from the Far East.

Thus far the essential economic context was similar though there were and are significant differences in the cost of coal production, the geological complexity of much German mining making the coal dearer to produce than its British counterpart. An even more significant difference was the overall position of the two national economies in the international energy market. The discovery of oil and gas in the North Sea provided Britain with a vital source of energy security, in which the economy as a whole might even benefit from instability in the international market, such as the oil crises of 1973 and 1979. The implications of those events for the German national economy were quite different: their vulnerability to fluctuations in the international energy market became evident. This became a motivation to slow down the contraction of coal and consider its retention at significant production levels as a kind of insurance against sudden reductions in the supply, or increases in the cost, of imported energy sources.

In strictly economic terms, then, it can be seen why German governments chose initially to make the contraction of the coal industry a gradual process, while British governments, especially after the privatisation of the steel and electricity supply industries, were prepared to sanction a sudden and severe reduction in the number of pits. The German holding operation was, however, undermined by further contraction in the steel industry and by European Community single market legislation, minimising government subsidies to staple industries.

However, though these important differences in the precise relationship to the contracting market may help to explain the differential rates of pit closure, they do not explain the very different approaches to how pit closure programmes should be handled. These are less attributable to economic than to political factors, themselves divisible into those of political structures and political ideologies.

The differences: political structures and ideologies

Formally, both countries are parliamentary democracies but the key difference lies in the extent to which power is centralised. Germany is, as its full title of Federal Republic implies, essentially a federal structure in which the Länder, or seats of regional government have considerable autonomy from and influence over the Federal Government. Britain is quite different. Though it has a long history of strong municipal government, this is generally at a very local level. The attempts to create a second tier of regional government has always, with the important exception of Wales, been limited and sporadic. Moreover, it has been an explicit part of Conservative Party policy over the last 15 years to assume very tight control over the functions and, most crucially, financing of local government. Especially because the major local authorities are in urban areas and frequently controlled by the Labour Party, the relationship between central and local government has been redefined as less one of mutual dependency or even creative tension than one of outright hostility. Hence a Conservative government will do anything in its power to prevent local Labour authorities being given increased capacity or resources of any kind, despite any regionally identifiable special needs – even, on one notorious occasion, funds from Europe's RECHAR programme, unless it could determine how the money should be spent. What little attempt there has been to develop regional policy has been designed to circumvent local authorities, at best allowing them some limited representation on quangoes under direct government control, the most obvious examples being the Urban Development Corporations. Thus even the most limited kind of potential co-ordination and planning, within the public sphere, has deliberately been rendered virtually impossible. Indeed periodically, the Government forces local authorities to bid against one another for public funding as in the City Challenge Programme.

One way of conceptualising this difference might be to suggest that in Germany a decentralised federal structure produces a particular way of handling contentious issues. Faced with possible conflicts of interest or political opposition, the main reaction seems to be to involve as many as possible in a forum, committee or other body, frequently given some power to recommend or undertake specific forms of action. In Britain, an increasingly centralised structure of government produces a quite different stance. Faced with conflicts or oppositions of interest, a clear decision is made by central government as to the correct policy, which is then imposed on the parties involved, with often only minimal consultation. In Germany, there seems to be what might be called a process of incorporation through institutionalisation, while in Britain the equivalent process is one of domination through exclusion. Each might be seen, in political terms, as a means of legitimising state power, but the processes involved appear diametrically opposed.

This is perhaps to overemphasise the difference, since many German contributors to this volume stress how the structures introduced to handle coal contraction were novel, even in the German context. Thus what we are witnessing is how a state seeks to resolve a new kind of problem. One political structure appears to lend itself to innovative policy responses while the other takes the form of conservative retrenchment.

Political structures alone do not wholly account for this difference. Prevailing political philosophies or ideologies also play a part. Again, there are superficial similarities. In recent times conservative parties have held electoral majorities in both countries. In common with the rest of

Europe, they have adopted free market philosophies, generally inimical to state intervention. However, this seems to have taken different forms in each case. In Britain, there has been a quite openly ideological attack upon a whole range of groups and institutions identified, in Mrs Thatcher's memorable phrase, as 'the enemy within'. Left-wing local councils, trade unions, liberal middle-class professions, groups whose activities or lifestyles symbolise dissent have all had their freedom for manoeuvre (in some cases quite literally) curtailed by government legislation and edict. This has involved quite specifically an attack on the ideas and assumptions of the post-war political consensus, especially around the goal of full employment and the provision of a welfare state. Hence political culture has been marked by a high degree of overt ideological conflict.

Germany seems to have held on more to an idea of 'consensus' albeit within a narrowly common set of assumptions around acceptance of free market capitalism. The differences, at least as they appear in the contributions to this volume, lie in the boundaries or limitations which are to be put around the operation of market forces. A phrase which seems to mark the difference, echoed by many of the German contributors, is that economic restructuring has to be undertaken in a way which is 'socially acceptable'. It is a measure of the difference between the two political cultures that such a definition is not accepted by the leaders of the British Conservative Party. Social considerations must not be allowed to interfere with the untrammelled exercise of market forces. Thus while German speakers at the conference took it for granted that unemployment was a social evil, in British political life it has become redefined as a (no doubt regrettable) necessity – 'a price worth paying' as one Chancellor of the Exchequer put it, in order to defeat inflation.

Even the occasional oblique references in German chapters to the perceived threat of electoral power and political violence in the Ruhr coalfield as galvanising Federal Government response can be seen as additional evidence of this contrast. In Britain mining areas can be safely regarded as peripheral to the political geography: The Conservative heartlands lie elsewhere. As response to the 1984–85 strike showed, the British Government was prepared to suppress picketing, even at the expense of civil liberties.

It is the combination of these factors which seems to me to best explain the differences in the way pit closures have been handled in the two contexts. It is not mainly a question of relative positions in the international energy market, or of differences in attitudes or willpower. It is, rather, a deeper difference, in the nature of state power: the way it has come to be habitually exercised and the checks and balances upon it.

Thus, as many of the German chapters show, there appears to be a vertical flow of communication and resources between Federal Government, the Länder and town or city councils. Correspondingly, there has been established a horizontal axis of equal importance, between public, private and voluntary sectors. The dynamic interplay of such interests and forces can produce at any level a consultative body which analyses, makes recommendations about and often implements policies affecting the nature of the local economy. In Britain, as has already been emphasised, vertical relations between central and local government are characterised by hostility, compounded by the absence of an effective regional tier of government. Any sense of horizontal co-operation is disabled by the mutual hostility of the public and private sectors, often exacerbated by government insistence on the predominance of the business community. This preferential treatment is evident in the wholesale transfer of power, especially at the local level, to quangoes (bodies set up with executive powers whose members are appointed by government), invariably with a majority of representatives from private industry. Thus any attempts at forming partnerships between public and private bodies either operate through quangoes which are already biased towards one set of interests, or operate at very local levels, often despite rather than because of the attitudes of central goverment. Hence, rather than any co-ordinated efforts to identify and resolve problems in the local economy induced by pit closure, as in Germany, the British response is piecemeal, fragmented and dependent largely on locally variable attempts at co-ordination.

Thus in Germany we find central, regional and local government working together on projects involving employers, trade unions and community groups. In North-Rhine Westphalia, Ruhrkohle plays a strategic role in the contraction of the industry and the regeneration of the local economy through diversification of its own operations. Extensive pre-redeployment training is linked into private and public networks of further education provision. Land reclamation is financially and legally systematic, with extensive planning controls over open-cast mining.

In Britain, by contrast, we find central and local government at odds over the closure programme and its effects, with no attempt to develop common policies. British Coal sees its priority as the creation of a profitable industry for privatisation, hiving off its social and economic responsibilities to the separate arm of British Coal Enterprise but otherwise accepting no role in the regeneration of local economies and with no interest in the diversification of its own activities. The retraining of miners is offered only after redundancy, haphazardly financed and undertaken outside existing further education provision. Land reclamation is at best uneven and open-cast mining is approved with minimum controls, even in the face of local opposition.

The cultural dimension

Hearing these papers as originally delivered at the conference and reading them now, it is easy to identify the positive examples to be drawn from the German experience but more difficult in the case of the British. The most obvious example is the adoption of Britain's coalition of local authorities to lobby on behalf of mining areas, the Coalfields Community Campaign, as a model for an equivalent group (ZAK) in Germany. Ironically, while the idea was British, it seems to have had more political impact on other European countries and on the EC itself.

There is also one other area where those involved with or concerned about mining communities in Britain demonstrate a sensitivity absent from the German contributions. This is around the themes of culture and community. Mining communities in Britain have traditionally been associated with a distinctive cultural pattern, in which the masculine nature of mining as an occupation, geographical isolation and a strong sense of collectivism have combined to make mining communities male-dominated, parochial and self-reliant. In particular, the nature of local institutions and kinship networks have provided mining communities with strong sources of support in times of adversity and effective forms of social control over aberrant behaviour.

The closure of the pit is thus seen to be more than a narrowly economic disaster, important though that is. It has also been seen as a threat to a whole way of life which cannot be sustained in the absence of a pit. It is not merely unemployment which has to be faced but the collapse of local political, welfare and cultural organisations and the corresponding network of informal relationships. The potential attenuation of community ties is seen as a loss in itself, measurable especially among young men, who, if they do not move away in search of work, may fall prey to various kinds of minor deviance, from petty theft to drug-taking. One powerful image of the social as opposed to economic effects of pit closure is of a rapid increase in the crime rate, disturbing and demoralising the community still further.

More widely, there is in Britain a more overtly cultural dimension to models of regeneration. There is simultaneously a sense that a cultural adaptation must accompany economic change and that there is much in the established culture to value and preserve. Ethnic pride makes this obvious in the case of Wales, since an integral part of national identity is at stake. Even within England, as in the case of Yorkshire, there is also a sense that regional identities and cultures are bound up with the fate of the coal industry and may not survive whatever follows it. This leads on to a stress on the need to invite local people into management of such economic and cultural change so they become the agents of change rather than simply being the victims of it. Perhaps, even in the North-Rhine Westphalian attempt at structural change, there is an excessive emphasis on the purely economic and a tendency to believe that such change can be induced from the top down by established institutions and organisations. The alternative might be to form new kinds of alliance and groups, through which those in ex-mining areas could themselves have a say and a stake in the regeneration of their own communities.

It remains true that it is an industry which has been removed and thus the provision of alternative means of livelihood must be the major priority. Yet this is not quite simply the equivalent of replacing the jobs from a factory which has been shut down. Economic change does not occur in a vacuum; it has consequences for family and community life. There is, for example, a general acceptance in the chapters in this book that vocational training should now become a life-long process but there are questions here about who, in terms of age and gender, should benefit from such training and, more broadly, how it relates to the education of the community as a whole. There is or should be a qualitative distinction between training provided from the outside for unemployed men and internally sustained education programmes for whole communities.

It is in part, to return to the guiding theme of this afterword, a problem of the relationship between local communities and the state. Current thinking on this question in Britain has been politically appropriated by the model of individuals as consumers of state services with the 'right to choose' implied by such a term, rather than by any active re-thinking of the idea of citizenship. The word 'community' has been redefined to disguise the return of financial responsibility for the sick and the infirm from the state to their liable relatives. 'Culture' has been appropriated as referring primarily to economic behaviour, as in 'enterprise culture' or the 'culture of dependency', marginalising questions about collective ways of living. Just as in Britain, the recent decline of the coal industry began with a fierce debate over what constituted 'economic viability', so perhaps the regeneration of ex-coalfield areas will go beyond the current German concern with what is 'socially acceptable', to a new agenda concerned also with what is 'culturally desirable'. But that would require a breadth of vision and imagination beyond anything visible in the responses of authority to coalfield decline and regeneration in either Britain or Germany.

INDEX